The Great
American Tomato Book

The Great American Tomato Book

THE ONE COMPLETE GUIDE TO GROWING AND USING TOMATOES EVERYWHERE

BY ROBERT HENDRICKSON

Doubleday & Company, Inc., Garden City, New York   1977

Library of Congress Cataloging in Publication Data

Hendrickson, Robert, 1933–
   The great American tomato book.

   1. Tomatoes.   I. Title.
SB349.H36   635′.642
ISBN 0-385-11437-0
Library of Congress Catalog Card Number 76–23767

FOR MY SISTER, CAROL—
MAY YOU GROW BETTER EVERY DAY

Tomatoes, tomatoes!
I hold them to view
Big ones, little ones
As fine as ere grew!

Old street vendor's cry

CONTENTS

INTRODUCTION

While protesting the pallid looks and vapid taste of commercially grown tomatoes, Representative James A. Burke of Massachusetts slipped one of the rocklike store-bought spheres from his pocket and dropped it so hard on the table that the sound may have awoken startled members of the Agricultural Subcommittee from their reveries of certain hot tomatoes. "It's like a baseball!" the congressman vociferated. "It's pulpy. . . . Doesn't even *taste* like a tomato. . . . Why don't we let the people find out what a *real* tomato tastes like?"

Bravo for Mr. Burke who was testifying in favor of a bill that would provide the public with free vegetable seeds, an undoubtedly meritorious proposal. His sentiments are those exactly of James Beard, who has written that "the most glorious of fruits, the tomato, has gone into a tragic decline . . . because it is being produced on a scale and in a manner that makes it an almost total gastronomical loss." Yet both congressman and cook might have been overlooking the fact that countless Americans actually *do* know what real, home-grown tomatoes taste like. In 1974, in fact, a record number of Americans—*close to half of all U.S. households*—grew vegetable gardens, and almost every plot contained a few tomato plants. For the first time in half a century, more garden space is being devoted to vegetables than flowers in America, and the vine-ripened tomato is without doubt the favorite of these vegetables. Four out of five people today prefer tomatoes to any other home-grown food, reports the U. S. Department of Agriculture, but really such has been the case for well over a century. The tempting tomato has in that time been used for more spicy sauces, canned in more soups, drunk in more juices, indispensable to more salads, slopped on more pizzas, grown in more home gardens, and pinched, poked, and haggled about in more markets than any ten of its closest competitors. No other fruit or vegetable has its mass appeal.

The truth is that only sweet corn (the second most popular home-grown vegetable) equals the tomato in taste when picked fresh from the garden. All vegetables are better fresh-picked, but fresh-picked corn and tomatoes taste like different species related only by name to celluloid, flannel-textured store-bought versions. Tomatoes are also the most versatile, prolific, and easiest to grow of all vegetables. They can be used as appetizers, desserts, main courses . . . can be juiced, stuffed, boiled, broiled, baked, stewed, fried, canned, frozen, pickled, and dried—or just eaten raw out in the garden with salt shaker in hand, a garden-ripe tomato still warm from the sun being a taste treat no one should miss. The fruits come in shapes, colors, and sizes that our ancestors wouldn't believe—including several *white* and *striped* varieties. They range in size from marblelike patio types to huge three-pound beefsteaks. They are round, ribbed, and elongated—hybridists have even developed a square tomato that when sliced fits neatly into sandwiches.

Tomatoes are low in calories, too, and rich in vitamins A, B_1, B_2, and C. There are non-acid types for ulcer sufferers, hollow types for stuffing, juicy kinds to make wine with. And to the great good luck of all, they grow almost anywhere—abundantly. Just one tomato plant will commonly produce twelve to twenty fruits, with yields of five times these amounts not at all rare. No wonder people are planting tomatoes everywhere—on front lawns, on patios and terraces, in rented vacant lots, on the "tar beaches," or roofs, of apartment houses, in pots and window boxes high up in skyscrapers, on ships and houseboats, even (in several cases) on the roofs of cars like Volkswagens. One man actually created a tomato garden on exactly one square foot of ground last year. He built a twenty-seven-foot earth-filled tower, drilled holes in the sides, and planted it with 230 tomato plants —which he had to harvest from an extension ladder!

This book is all about the tomato, *solely* about the tomato, its aim to make the best growing techniques, preserving methods, and tomato recipes available all in one volume to both experienced vegetable gardeners and the 2.5 million beginners who join their ranks every year. With that in mind, and the belief that all gardeners like to know what they grow, we'll begin this planting guide with the only complete history of the tempting tomato to be found anywhere. So here for your delectation, replete with further panegyrics, is . . .

The Great
American Tomato Book

The wild Peruvian tomato—the progenitor of all modern tomato varieties.

Chapter One

THE AMAZING XTOMATL...
OR LOVE APPLE...OR MAD APPLE...
OR RAGE APPLE...
OR WOLF PEACH...
AMERICA'S FAVORITE VEGETABLE...
OR FRUIT...OR BERRY...

While the delectable tomato makes us wish we had double our 5,500 taste buds, it did not always do so. *Lycopersicon esculentum* var. *commune* surely ranks among the most interesting, mouth-watering, taste bud-tingling treats of all time, but what is now the most popular fruit of the vegetable vale was once decidedly the most reviled. For the tomato is the infamous "love apple" or "wolf peach" of antiquity that our Puritan progenitors considered a forbidden fruit, the "mad apple" or "rage apple" which our forefathers feared as a poisonous plant. It took more than three hundred years for the outcast tomato, which was discovered here in the New World, to be accepted back in its ancestral home.

L. esculentum (even the tomato's scientific patronym sounds delicious) is distinctly a New World plant native to South America, where it first grew in the Andes and is still found in its wild, wrinkled, and wizened state throughout Peru, Bolivia, and Ecuador—a tiny fruit no bigger than a marble that grows in clusters on a small, sparse vine. Like corn, potatoes, peanuts, the pumpkin, tobacco, and Bourbon, the Elysian food is an American gift to the world. It was first cultivated in Central America and soon introduced into Mexico, but was probably never grown by North American Indians because of its sensitivity to cold. The Mayans, whose seafaring traders brought tomato seed to Yucatán, called the then tiny fruit *tomatl,* or *xtomatl,* and so prized *L. esculentum* that they traced its form on their pottery. When Cortez and his band escaped the first Aztec uprising in Mexico, they managed to buy tomatl seeds in the great market of Chichén Itzá to bring back with their plunder to Europe, where the plant was initially called both *pomi del Peru* and *mala peruviane*.

At first *L. esculentum* met no resistance in Europe, generally living up to its name and enjoying a reputation as an esculent dish indeed. The Spanish, along with other Mediterranean peoples, widely adopted the tomato and when an ingenious Spanish chef combined the fruit with olive oil, spices, and onions, he created the world's first tomato sauce, which was hailed by the Spanish court. Soon, as was inevitable with every new and rare food, *L. esculentum* even acquired a reputation as an aphrodisiac.

Why tomatoes were dubbed "love apples" is a matter of some dispute. For one thing, the tomato fruit was said to resemble the human heart (the seat of love, according to the ancients) and thus by the doctrine of resemblances was an aphrodisiac. That tomatoes hailed from exotic climes and were a shapely, scarlet fruit also undoubtedly helped, but the designation owes just as much to semantics as sexuality. All Spaniards at the time were called Moors and one story has it that an Italian gentleman told a visiting Frenchman that the tomatoes he had been served were *"Pomi dei Moro"* (Moor's apples), which to his guest sounded like *"pommes d' amour,"* or "apples of love." However, another version of this old yarn claims that "apples of love" derives in a similar roundabout way from the Italian *pomo d' oro* (golden apple), which is identical to today's Italian name for the tomato, *pomodoro*. This last is possible because yellow or golden tomatoes were among the first varieties to be introduced to Europe.

In any event, the tomato quickly gained a reputation as a wicked aphrodisiac, and justly or not, held this distinction for many years. In Germany, its common name is still *Liebesapfel* or "love apple." As for the word's pronunciation, those who, like the English, pronounce it the "toe-mah-toe" are probably historically correct. The fruit was first called *tomate* in Spain and Portugal and pronounced in three syllables. The final "o," incidentally, has no place at all in "tomato," apparently being there because mid-eighteenth-century Englishmen erroneously believed that it should have this common Spanish ending.

The tomato's scarlet past probably contributed in part to its notoriety as a poisonous plant: *Lycopersicon esculentum* was reputedly a *deadly* aphrodisiac. This is reflected in the tomato's first botanical name, *Lycopersicon,* which means "wolf peach" in Greek, the fruit's full scientific name translating as "juicy wolf peach." At any rate, the noted English traveler John Gerard wrote in his sixteenth-century *Herbal* that "these Apples of Love . . . yield very little nourishment to the body and the same naught and corrupt." Similarly, the horticulturist Philip

Miller, a colleague of the great Linnaeus, observed that there were those who thought that "the nourishment [tomatoes] afford must be bad." Most likely, the tomato acquired this stigma because it is a member of the deadly nightshade family, which belongs only in Dr. Rappaccini's garden. As early as 1544, the Italian herbalist Pietro Andrea Mattioli, in his *Commentaries on the Six Books of Dioscorides,* linked the "golden apple" with mandrake, henbane, and belladonna, all extremely poisonous plants.

It has since been established that all parts of the tomato plant *except* the fruits are toxic, containing dangerous alkaloids, and to this day people are caused severe digestive upset from eating them. The tomato belongs to the Solanaceae family (also called the nightshade, or tobacco or potato, family), which includes perhaps three thousand species—tobacco, narcotics, and many flowers and vegetables among them. It may well be that the tomato suffered, on a smaller scale, the fate of its cousin the potato, which was introduced to Europe at about the same time. The first potato plants from America were presented to Queen Elizabeth by Sir Walter Raleigh, and her chamberlain soon planted them along the Thames. He then invited the local gentry to a banquet featuring potatoes at every course. But instead of cooking the tubers, the unsuspecting chamberlain cooked the poisonous stems and leaves. A mass stomach-ache resulted that set back the acceptance of potatoes in England some two hundred years. Likewise, tomatoes were for centuries mostly cultivated in English greenhouses as a floral ornament. Ketchup, however, was made from the juice of tomatoes in Old Blighty all along, the first O.E.D. source for the sauce date being 1711.

Empress Eugénie, Napoleon III's wife, introduced Spanish tomato dishes into France in the mid-nineteenth century, just as she introduced a number of fashions that held sway at the time. This beautiful, elegant, and charming woman, the undisputed leader of French fashion, died in 1920, ninety-four years old and a legend in her time. But whether or not tomato dishes made Eugénie love and live longer, she cannot claim the distinction of being the *first* to introduce the fruit itself to the French cuisine. This honor must go to Napoleon's chef, who invented Chicken Marengo at the Battle of Marengo in Lombardy during the Italian campaign, the quite exact date being two o'clock in the afternoon, June 14, 1800. Having no butter on hand, the resourceful cook sautéed the chicken in the local olive oil, adding the sauce consisting of tomatoes, herbs, mushrooms, and Marsala wine that is still called Chicken Marengo. Later Antonin Carême—the starving child of a family of twenty-five who became "the chef of Kings and the King of chefs"—featured stuffed tomatoes as his favorite accompaniment for a filet of beef. Carême would stuff tomatoes *à la Sicilienne,* with ham, chopped onions, etc.; *à la Florentine,* using chicken and other ingredients; and *à la Provençale,* employing chopped mushrooms, garlic, and bread crumbs in his stuffing. It wasn't long before all of France was influenced by "Carême of Paris," as Louis XVIII granted him the right to call himself, and by the time Empress Eugénie came along, the tomato was already on its way to becoming almost as indispensable to the French cookery as it is to the Italian, all French *Provençale* or *à la Portugaise* dishes being made with them.

Unfortunately, America went along with England and while southern Europe and France were enjoying their savory tomato sauces, *L. esculentum* was grown here only as a curiosity or as an attractive ornamental trained on trellises. The fruit was condemned from the pulpit and marked with skull and crossbones by the doctors. Thomas Jefferson, who once wrote that he would rather be a common dirt gardener than President, did grow tomatoes at Monticello in 1781; they were featured in George Washington's gardens at Mount Vernon as an ornamental plant; the New Orleans market offered them to more knowing housewives of French descent as early as 1812; and foreign visitors and Yankee sea captains introduced tomatoes to New England and Philadelphia toward the end of the eighteenth century (housewives using them to make "kechap" or "ketchup"). But these were barely noteworthy exceptions. The tomato was for the most part still guilty by association, and far too scarlet and shapely for the Puritan palate. In Salem, Massachusetts, for instance, ancestors of the original witch-hunters wouldn't touch the tomato with a ten-foot fork as late as 1800, although its juice was "applied externally to remove eruptions upon the skin."

If any one person liberated *L. esculentum,* it was Colonel Robert Gibbon Johnson, an eccentric gentleman of Salem, New Jersey. In 1808, after a trip abroad, Johnson introduced the tomato to the farmers of Salem, and each year thereafter offered a prize for the largest locally grown fruit. But the colonel was a forceful individualist and wanted his introduction to be regarded as more than an ornamental bush. One day in 1820 (some say 1830), he announced that he would appear on the Salem Courthouse steps and eat not one but a whole basket of "wolf peaches!"

Public reaction in Salem was immediate. Johnson, forty-nine years old, an impressive-looking man with high forehead, long hooked nose, powerful chin, and iron-gray hair, was widely known as Salem's first citizen, a member of one of the community's pioneer families. But of his eccentricity, there was no doubt. Just look at the dress the man affected, people said; why he aped his friend George Washington, still wore a black suit with white ruffles, black shoes, tricorn hat, black gloves, and cane! Even copied the general's mannerisms. True, he was a man of his own mind. During the Revolutionary War, when he was all of seven, Johnson had slapped a British officer in the face. Later he had marched in the

Whiskey Rebellion. And after he'd been accidentally locked out of the Episcopal Church one Sunday, he changed his religion in a huff, joining the Presbyterians and giving them the land next to his mansion to build a church of their own.

But this was entirely too much! Eating the deadly wolf peach or Jerusalem apple—a whole basket of the vile Things! Even if he did claim they were "used as a food by the Egyptians and Greeks, only to be lost in the annals of antiquity." Why that crazy colonel was courting certain death! Declared Johnson's physician, Dr. James Van Meeter: "The foolish colonel will foam and froth at the mouth and double over with appendicitis. All that oxalic acid! One dose and you're dead. Johnson suffers from high blood pressure, too. That deadly juice will aggravate the condition. If the wolf peach is too ripe and warmed by the sun, he'll be exposing himself to brain fever. Should he survive, by some unlikely chance, I must remind that the skin of the *Solanum lycopersicum* [as it was then called] will stick to the lining of his stomach and cause cancer. . . ."

High noon for the tomato came on September 26, 1820, and Dr. Van Meeter was there, black bag in hand, along with two thousand other curious people from miles around, to watch Colonel Johnson commit certain suicide. Johnson, an imposing figure dressed in his usual black suit and tricornered hat, solemnly ascended the courthouse steps as the local fireman's band played a dirgelike tune. Selecting a tomato from his basket, he held it aloft and launched into his spiel: "The time will come when this luscious, golden apple, rich in nutritive value, a delight to the eye, a joy to the palate, whether fried, baked, or eaten raw, will form the foundation of a great garden industry, and will be recognized, eaten, and enjoyed as an edible food. . . . And to help speed that enlightened day, to help dispel the tall tales, the fantastic fables that you have been hearing about the thing, to show you that it is not poisonous, that it will not strike you dead, I am going to eat one right now!"

Colonel Johnson bit and his juicy bite could be heard through the silence, until he bit again, and again, and again—at least one spectator screaming and fainting with each succeeding chomp. The crowd was amazed to see the courageous colonel still on his feet as he devoured tomato after tomato. "He's done it!" people shouted. "Look at him; he's still on his feet eating the blamed things; he's still alive!" Johnson soon converted most onlookers, but not until the entire basket was empty did Dr. Van Meeter slink away and the band strike up a victory march and the crowd begin to cheer.

Colonel Johnson's bite was heard around the country, if not the world. His efforts at least turned the tide for the tomato (though he was not, of course, the first American to eat the fruit) and it was regularly appearing in markets across the land by 1835, when the editor of the *Maine Farmer* wrote that tomatoes were "a useful article of diet and should be found on everyone's table." Nevertheless, prejudices still lingered. As late as 1860, the popular *Godey's Lady's Book* warned its readers that tomatoes "should always be cooked for *three hours* before eating," and word detective Dr. Charles Funk noted in his *Horsefeathers* that at the turn of the century, in rural Ohio, his mother was averse to eating what was then known "only as the love apple and believed to possess aphrodisiac properties, and was therefore feared by virtuous maidens." The myth still persists that tomatoes make the blood acid and a "health food" brochure printed as recently as 1970 warns that: "Many vegetables are actually toxic and consuming them will pollute the body and taint the brain. Tomatoes are a bigger threat to the health of this country than any other vegetable because they are related to such poisonous plants like [sic] nightshade, belladona, and tobacco."

Actually, there isn't much meaningful controversy about whether the tomato is poisonous anymore, and not many people argue about whether it is a fruit or a vegetable. Legally, *L. esculentum* is generally a vegetable, and botanically it is a fruit or berry belonging to the potato family. This matter was settled in 1893 when an importer contended that the tomato was a fruit and therefore not subject to vegetable import duties. The United States Supreme Court held that the tomato (and all plants, including potatoes, cabbage, carrots, peas, celery, and lettuce) had to be considered a vegetable when it was served in soup or with the main course of a meal, although it could be considered a fruit when eaten out of hand or as dessert. The scientists, of course, did not consider this legal reasoning very scientific and so we are stuck with the inconsistency today.

Tomatoes are presently a big business in the United States, with some 450,000 acres allotted to them throughout the country. Acreage devoted to tomato farms has drastically declined over the last decade, but improved tomato varieties have actually increased production. Domestic output exceeds 1,013,000 tons of fresh tomatoes annually, and another 376,000 tons are imported into the country. Additionally, nearly six million tons of tomatoes are processed into canned goods each year. Unfortunately the billion dollar crop is too often harvested by migrant "stoop laborers," who suffer squalid living conditions and are paid a penny a pound—another good reason to grow your own.

The tomato has become so popular that it is part of American folklore. The Campbell Tomato Soup can, for example, has with the advent of Pop Art became as traditional as the Old Man of the Mountain, Andy Warhol's painting of it having last sold for $60,000—quite an advance from that day in 1897 when a young $7.59 a week chemist named John Dorrance worked out the formula for condensing it and Campbell's Tomato Soup was born.

Tomatoes figure prominently in contemporary literature, too. Some readers will remember that Don Correlone in the *Godfather* died a relatively peaceful death while tending his tomato patch. For that matter, the tomato has even had a popular science fiction story written about it. "Rogue Tomato," by Michael Bishop (which can be found in Harper & Row's *New Dimensions, 5; Science Fiction,* 1975), tells of a man who wakes one morning to find himself a limbless, planet-sized tomato somewhere in outer space. Living in constant fear of being eaten by giant tomato worms or being stewed by the heat of the sun, he contemplates ending his life—but how can a tomato commit suicide? Readers will have to buy the book to find out the ending.

If there is controversy about the tomato in modern times—aside from minor disputes such as whether clam chowder should be made with milk or tomatoes (see Chapter Eighteen), which can become violent among gourmets—that controversy concerns their taste. There are those who argue that many Americans never know the indescribable taste of a sun-ripened tomato, and judging by store-bought varieties, they are absolutely right. The same applies to the authenticity of many tomato products. One leading tomato catsup manufacturer, for example, recently spent millions developing a fresher, richer, more natural tasting product. Sadly enough, the new catsup was rejected by consumers and the firm had to go back to its old overcooked sauce—sales returning to normal when it did so. Fresh tomato flavor, concluded a company official, just ain't familiar anymore.

Today's commercial tomatoes are grown for toughness and uniformity of size, which makes things easier for harvesting and processing machines; for tough skins that will ship and store well; and for yield at the expense of flavor. Taste is hardly ever considered. In fact, most fresh tomatoes are picked green or light pink for market because the fruit is perishable when ripe and ripens even under refrigeration. Tomatoes are therefore ripened by ethylene gas after being shipped, and although ethylene is harmless—a gas naturally given off by bananas and McIntosh apples—this artificial process deprives the tomato of the vitamin C it would normally gain from sunlight and substantially subtracts from its flavor.

Commercial tomatoes quite often taste like their plastic packages. In order to get any pleasure at all from store-bought specimens, they must be wrapped in newspaper until they fully ripen and then scalded and peeled before being eaten. Much preferable is the old salt-shaker method cited by Joan Faust, garden editor of the New York *Times*. "This is the one made around mid-July when the patient gardener can't wait any longer," Miss Faust writes. "He ventures out to the vegetable patch, salt shaker in hand, to pluck just one rosy tomato to see how things are coming along. Who can describe his ecstacy when biting into the first tomato, lightly dusted with a dash of salt. Let's hope that this heritage is never lost to the next generation and that children will not grow up thinking tomatoes come from plastic-wrapped cardboard trays. . . ."

Fortunately, the statistics show that no matter how you slice it, the scarlet, garden-ripened tomato with the unsavory past and savory taste will never be forsaken. Whether it is eaten Miss Faust's way or in way-out dishes like the "whipped cream à la tomate sprinkled with brandy" invented by futurist chef Jules Maincave, or as candied tomatoes, or even in something called tomato sherbet, it's safe to say that the home-grown tomato will remain an American institu-

For better or worse, the progeny of those tiny Peruvian tomatoes —truckloads of modern, commercially grown tomatoes.

tion. "Tomatoes can be prepared in so many delicious ways that one can eat them every day and not get tired of them," George Washington Carver wrote. Whether used in salads, spaghetti sauce, marmalades, minced meats, or a hundred other dishes, the tempting tomato will always be first choice for the backyard patch or those window boxes high in the sky, the first fruit (or vegetable) of the American table. The only real problem for people will be which of the hundreds of tomato varieties to grow, and exactly where and how to grow 'em. . . .

Chapter Two

THEIR INFINITE VARIETY:
CHOOSING THE BEST LOVE APPLE FOR YOU

There are so many tomato varieties available from seedsmen today, well over five hundred, that you probably couldn't care less that the cannibal's tomato can't be purchased anywhere—unless some Fiji Islands cannibals are among your acquaintances. The cannibal's tomato, *Solanum anthropophagorum* (its botanical surname the classical term for man-eating), resembles the smaller cherry tomatoes we eat except that it is two-lobed with a nipple at the apex. Native to the Fiji Islands, the fruit of the species was formerly (maybe "is presently") eaten by cannibals as a unique "chaser." If you can get hold of seed from a rare-plant

Burgess Stuffing Tomato—a hollow type bred especially for those who like to stuff and bake tomatoes.

dealer, the cannibal's tomato is easily grown in the greenhouse, but we're told it's not much good without the main course it usually enhances.

Modern tomato varieties started proliferating in America in about 1895, when scientific breeding practices began improving fruit quality and developed disease-resistant types and hybrids of superior size for the first time. Today the U. S. Department of Agriculture alone tests two thousand varieties of tomatoes some years at their experiment stations throughout the country, and plant breeders at universities and seed companies evaluate thousands more. New types from all over the world are extensively tested for resistance to scores of insects and diseases, for nutritional value, and for productivity. Taste is also considered, but this is at

Tomatoes come in an almost infinite variety of shapes and sizes.

Infinite Variety: Choosing the Best Love Apple for You / 11

best a subjective matter and is all too often made secondary to the shipping and storing qualities so important to commercial growers.

Most home gardeners have no interest whatsoever in "firm-skinned" or "good shipping" tomatoes, which can be bought in any supermarket, and when reading seed catalogs should obviously consider all the other qualities cited. These are myriad and many are described in the pages following. Size, shape, color, mildness (acidity or non-acidity), and cracking-resistance are all important factors to evaluate when selecting home varieties. Height and bushiness of the plant are other serious considerations—especially for anyone intending to grow patio or window-box tomatoes. Your goal may also be the best tomatoes for preserving or making tomato paste. Disease resistance may be of particular interest because a particular disease is prevalent in your area, or you may be concerned about the "number of days to maturity" (the time it takes for a plant to bear fruit) because the growing season is short where you live. On the other hand, it might be that you're primarily interested in choosing a truly unique plant, a novelty different than anyone in the neighborhood grows—and there is an abundance of unusual tomato types, ranging from tree tomatoes to white, green, and striped varieties.

The point is that growing tomatoes from seed opens up a new world of possibilities. It is easier to begin with potted or bare-root plants bought in a local nursery (see Chapter Four), but you'll be missing one of the great joys of tomato gardening in not raising plants from seed. Perhaps even more importantly, no nursery could possibly make available the wide selection of tomato varieties offered in seed catalogs (Burpee alone lists 35 varieties, Burgess 79, and Stokes 95). And while hometown nurseries usually stock varieties that do well locally from both bearing and disease-resistance standpoints, their plants are often poorly grown. Finally, too many unscrupulous nurserymen cash in on famous names, labeling a plant a Burpee Big Boy, for example, when it is nothing of the kind—and there is no sure way for a gardener without a Ph.D. in botany and a high-powered microscope to tell the difference between varieties until a tomato plant bears fruit.

When you buy tomato seed from a seedsman, don't worry about getting enough. One packet contains up to a hundred seeds (there are five thousand tomato seeds to the ounce), far more than even the largest family will need. Six to twelve plants, if properly grown, should supply a family of five with more than enough tomatoes, and any seed left over can be sealed and stored in a dark cool place for up to five years without losing much viability.

It is wise to deal with a seed company that has its growing fields in a region with climatic conditions similar to your own. (See Appendix III.) In any event, read all of the information on the seed pack carefully, taking care to see that the date on the pack coincides with the year in which you are planting. (Don't buy a 1976 pack for planting in 1977.) As for preplanted seed tapes, these may make planting a mite easier as respects spacing and thinning plants, but they are not available in many varieties, do not always germinate well, are hard to work with when wet, and are much more expensive than packs of seed. You'll have so few seeds to plant that tapes just aren't worth the extra money. The preplanted seed flats sold everywhere today are another story and have more to recommend them, but they, too, come in fewer varieties and cost much more.

A wide selection of the best and most unusual tomato varieties is given in this chapter, but keep in mind that no tomato variety can transcend all others in every respect—varieties are bred with particular traits in mind and usually excel in these. Don't be disappointed if a particular variety doesn't perform quite as well as the gardening catalogs claim, either—like the tomato itself, catalogs are best digested with a few grains of salt.

The best general advice is to try four or five of the nearly four hundred varieties described in these pages. Also subscribe to several of the many seed catalogs listed in Appendix IV, for new, improved tomatoes are developed every year. When reading a catalog don't pay as much attention to the pictures of perfectly grown fruit as you do the fine print describing the plant. Days to harvest and disease resistance are particularly important to know, as we'll see, and any tomato described in words like "for the home garden" is usually a recommendation of good eating quality.

Stay away from outright novelties if you're a beginner and of course totally ignore "market varieties" bred for thick skins and similar qualities. Do this for a few years and you'll surely find the tomato with the right flavor for you. Just be certain to include among your first choices at least one plant each of an early, midseason, and late ripening variety, so that there will be fresh tomatoes on the table throughout the gardening season. Two complete seasonal lists, the most thorough of their kind available, can be found in Appendixes I and II at the end of this book. These master lists also contain specific information about all the love apples following that you can choose from—many of which will seem just as unlikely as the "cannibal's tomato."

SELECTING THE SUPERTOMATOES

The most colossal tomato ever grown turned crimson with the sun one August morning in Calaveras County, California, back in 1893. So at least gardening writer Edward J. Wickson claimed at the time, when he described a Brobdingnagian specimen measured at: "circumference, 22½ inches; diameter 8 inches; and weight 4½ pounds." It will be hard to beat that record, which has stood for

One of the supertomatoes—Ponderosa, which often weighs two pounds and more.

Ultra Boy—one slice of these giant fruits makes a sandwich.

nearly a century, but many tomatoes do develop into Bunyonesque fruits when properly grown. If you're out to make the *Guinness Book of World Records,* it is best to start with one of the varieties below, which can grow over three pounds in weight. For other characteristics check these supertomatoes against Appendix II:

Abraham Lincoln

Beefeater

Beefmaster Hybrid

Beefsteak

Big Crop Climbing

Boatman Miracle Climber

Brimmer

Burgess Jumbo Hybrid

Burpee Big Boy

Climbing Trip-L-Crop

Colossal

Crimson Giant

Dutchman	Pink Ponderosa
Garden Master	Pondeheart
Giant Belgian	Ramapo
Giant Oxheart	Superman
Giant Tree	Super Master
Globelle	Supersonic
Golden Boy	Terrific
Golden Oxheart	Ultra Boy
Jung's Giant Climber	Vineripe
Oxheart	Watermelon Beefsteak
Park's Whopper	Winsall
Pink-skinned Jumbo	Wonder Boy

MIDGET TOMATOES

The littlest tomatoes available range in shape from cherry and plum types to tiny tomatoes that look like eggs. They grow on plants as small as a foot in height. These miniatures are discussed fully further on in relation to indoor, window box, hanging basket, patio, and small-space gardens. Size of both plant and

Some of the many midget tomatoes in a salad with both eye and taste appeal.

fruit were considered in selecting the varieties recommended here. Often these midgets will bear up to two hundred little fruits on a single plant:

Basket Pak	Red Currant
Burgess Early Salad	Red Peach
Droplet	Red Pear
Dwarf Champion	Red Plum
Early Red Cherry	Saladette
Early Salad	Small Fry
Epoch Dwarf Bush	Stokes Alaska
German Dwarf Bush Imp	Subarctic
Grape Tomato	Sugar Lump
Hybrid Pickle	Sugar Red
Johnny Jumpup	Sugar Yellow
Merit	Tiny Tim
Patio	Tom Tom
Pixie Hybrid	Toy Boy
Porter	Tumbling Tom
Presto	Window Box
Pretty Patio	Yellow Tiny Tim
Red Cherry	

"TOMATO FACTORIES"

Some modern varieties are so prolific that they are virtual "tomato factories." A number of tomato plants actually grow more than twenty feet high when trained against a building, and several gardeners claim to have harvested 250 tomatoes from a single rampant vine. There is documentation of a sprawling, un-staked California plant that "covered a space of eight feet *square*" and yielded 170 pounds of fruit before frost killed the vine. Many of the preceding "Super-tomatoes" are capable of such heavy fruiting, with proper watering and fertilization, but below are the best plants for gardeners out to break production records:

Boatman Miracle Climber	Ponderosa
Climbing-Trip-L-Crop	Terrific
Giant Tree	Trellis 22
Jung's Giant Climber	Vineripe
Lakeland Climber	Winsall
Oxheart	

Climbing tomatoes grow up to thirty feet high, producing baskets of fruit.

Rutgers, a prolific, delicious old-time favorite.

Almost every tomato mentioned in these pages is a gourmet favorite; you'll have to experiment to find the one with exactly the right taste for you. Scientifically, tomato flavor depends to a large extent on the ratio of sugar to acid. If the ratio is too low, the fruit will be sour and have an insipid taste. Usually, the greater the percentage of jelly and seeds, as compared to flesh, the lower the sugar-acid ratio. So say the experts, opting for a fleshier tomato. But many people prefer a tomato packed with seeds and for this reason grow large-fruited varieties with five to ten seed chambers. (Small tomatoes have only two chambers.) And if fleshiness is so important, why aren't paste-type canning tomatoes the American favorites for eating out of hand? There's just no explaining tastes— except to say that everyone prefers a tomato ripened on the vine.

DELICIOUS TOMATOES

Infinite Variety: Choosing the Best Love Apple for You / 17

NUTRITIOUS TOMATOES

All tomatoes are known as "the oranges of the vegetable garden" for their high vitamin C content. All are rich, too, in vitamins, A, B$_1$, and B$_2$ and contain only about four calories per ounce. It is also interesting to note that any home-grown tomato ripened on the vine has one third more vitamin C than a commercially grown, artificially ripened fruit, a fact firmly established in studies made at Michigan State University's Department of Food Science. Nevertheless, a handful of tomato varieties are special developments that are especially nutritious. Try Burpee's Jubilee; Caro-Red; Doublerich; Peron; and many of the oranges, yellows, blues, and pinks described farther on. Doublerich, which was developed by crossing garden types with the tiny wild Peruvian tomato, has twice the vitamin C of most tomatoes. Caro-Red, a flavorful orange type, is ten times richer in vitamin A than ordinary varieties due to beta-carotene, its orange pigment. One Caro-Red tomato provides up to two times an adult's minimum daily requirement of vitamin A and is rich in vitamin C as well.

Orange-fruited Jubilee is low in acid, high in vitamins.

More brilliant than any other tomato is Trimson, a variety recently developed from a wild Filipino species by scientists at the University of Toronto. Trimson has the most intense red color of all tomatoes, two or three times the red color intensity of common red types. Crimson Sprinter and Moira, two similar varieties in the high crimson series, are also intensely red. Among ordinary tomatoes Red Glow, Big Red, and Early Red Chief are brighter than most.

THE REDDEST TOMATOES

The attractive orange and yellow tomatoes are often a little higher in vitamin C and lower in acid content than their red relatives, making them ideal for those with stomach disorders. They can be large slicing tomatoes or small cherry types that are early or late in fruiting. Old favorites include: Jubilee; Giant Oxheart; Golden Boy; Golden Delight; Golden Jubilee; Golden Oxheart; Golden Ponderosa; Golden Queen; Golden Sphere; Jubilee; Mandarin Cross; Moon Glow; Morden; Orange Queen; Sugar Yellow; Sunray; Yellow Pear; Yellow Plum; and Yellow Tiny Tim. Of tangerine tomatoes, there is but one—an "intensely tangerine-colored" variety called Tangella that is only offered by England's Sutton and Sons. (See Appendix IV.)

ORANGE, YELLOW, AND TANGERINE TOMATOES

When pink tomatoes are a purplish-pink, they are often called "blue." The pinks, which are second lowest in acid content behind orange and yellow varieties, are nevertheless very similar in flavor to reds. They, too, can be early, main crop or late, little cherry types or giant beefsteaks. Especially recommended are: Beefmaster; Belgian Giant; Burpee's Globe; Crackproof Pink; Dutchman; Dwarf Champion; Early Detroit; Early Pink; Gulf State; Kurihara; Livingston Globe; Laketa; Mission Dyke; Oxheart; Pink Deal; Pink Gourmet; Pink Lady; Pink-skinned Jumbo; Ponderosa Pink; Pondeheart; Watermelon Beefsteak; and Winsall.

PINK OR BLUE TOMATOES

Yes, Virginia, there is a white tomato variety—several of them, in fact. Like pink-fruited and orange types, the whites are relatively low in acid, but they are a pale color very close to white when ripe, which makes for good conversation as well as good eating. Their flesh is close to paper-white in color. Whites take from seventy to eighty days from transplant to fruiting. Good varieties of this average-sized tomato include Snowball, White Wonder, White Queen, and White Beauty.

WHITE TOMATOES

White Beauty, one of the few pure-white tomatoes—ivory white on the inside and outside when fully ripe.

GREEN TOMATOES

Luther Burbank would be green with envy, but the green tomato, another relatively new introduction, is nevertheless *green when fully ripe*. Perhaps this detracts from its taste, if the eye and taste buds work in tandem, as gastronomes tell us they do. Yet green tomatoes are surely a conversation piece, or something to win a bet on. ("I say that this tomato is perfectly ripe!") One such variety is Evergreen. Low in acid, Evergreen is a firm, sweet fruit good to eat out of hand. It "ripens" in seventy to seventy-five days. Maritimer, an earlier green, ripens in fifty-nine days and is often used for canning.

STRIPED TOMATOES

Tomatoes with distinct stripes are probably the ultimate in tomato novelties. "Tiger Tom," which has attractive stripes in red and orange-yellow, and "Tigerella," which has red skin with golden stripes, are available only from Sutton's in England. They are described as early-ripening, medium-size fruits, but I have grown both here and find that the tomatoes are about the size of a large plum tomato, though round.

ODD-SHAPED TOMATOES

By no means do tomatoes have to be round—some of them refuse to conform. The old favorite Pink Ponderosa often sets fruits that are grotesquely shaped (to the conformist), and old-fashioned ribbed types such as Beefsteak mentioned in

Tangella tomatoes (right) and the striped variety, Tigerella. The stripes on Tigerella are more pronounced on smaller-sized fruits.

the master lists also offer a striking contrast to today's uniformly round varieties. Liberty Bell, a new bicentennial-year introduction, is actually shaped like a bell pepper; Sausage, too, is appropriately named; and Square Tomato, like several paste types, is almost square in form (good for sandwiches!). Try also some of the small-fruited varieties, such as egg tomatoes, or Red Peach, which is vase-shaped.

Petomech II, one of the newest of the so-called "square" tomatoes that are good for processing. Other varieties have an even more pronounced square shape.

UNIQUE TOMATO FOLIAGE

Tomato plant foliage isn't always green. Abraham Lincoln, for example, has bronze-colored leaves that would stand out as a background in any flower garden. As for leaf shape, Giant Italian Tree sports enormous potatolike leaves. The perennial Tree Tomato has leaves over a foot long and fragrant purple and green flowers, while the Husk Tomato bears its small fruit all wrapped up in brownish parchment.

The Indians of ancient Peru first cultivated the perennial tree tomato, often on mountainsides at elevations up to eight thousand feet. *Cyphomadra betacea,* as it is called botanically, is a member of the Solanaceae family and therefore related to the garden tomato. Once established (usually in two years), this elephant-eared plant bears fruit up to seven months a year, eventually reaching a height of eight to twelve feet. Propagation is generally from seed, which germinates easily, but plants can also be purchased. The tree tomato can be grown outdoors all year round in warmer climates where the temperature doesn't fall much below 50° F., but must be taken in in the North, for the winter months, and grown in the greenhouse or a large sunny window. Transplant a seedling plant to a small pot the first year, increasing the size of the pot each year to accommodate its roots until the plant finally occupies a large pot. Tree tomatoes, which are very ornamental, should be grown in rich loam, with leaf mold and well-rotted manure added to it. The ripe fruit is orange-red, smooth-skinned and egg-shaped, hanging in clusters. Sweeter in taste

A full-grown tree tomato; notice the egg-shaped tomatoes it produces.

than the tomato, it is subacid and can be eaten raw for a dessert or used for jams, preserves, or stewing. The plant is usually cut back in winter to encourage spring flowering.

SWEET HUSK
TOMATOES
(GROUND CHERRIES)

These worthwhile novelties, which belong to the same family as the garden tomato, have been in cultivation in the warm countries of the world for over two centuries. Hawaii's renowned Poha Jam is made from husk tomatoes. The fruits are also dried in sugar and used like raisins. Several species are grown in America for their yellowish, cherry-sized fruit, which is enclosed in a parchmentlike husk, can be eaten raw, and makes excellent pies and preserves:

1) The Ground Cherry or Strawberry Tomato (*Physalis pruinosa*) resembles a dwarf outdoor bush tomato, usually growing no more than eighteen to twenty-four inches tall. An annual, it can be bought nursery-grown or started from seed about the middle of March and set outside ten weeks later. Follow cultural practices for the garden tomato and these plants will bear prolifically. Their heavy husks are easily opened to reveal deep yellow fruits (when ripe) which have a sweet flavor that blends well with other fruits.

2) The Tomatillo or Jamberry (*Physalis ixocarpa*), another annual, is cultivated exactly like the ground cherry. A taller plant, three to four feet high, it yields a larger fruit—up to an inch in diameter—that is used in the same way. One difference between the two species is that tomatillo fruit completely fills its husk and often bursts revealing the ripe fruit. As a result, it is more difficult to remove the sticky fruit from the husk, but this can be done by soaking the husk

These husk tomatoes or ground cherries, all wrapped up like presents in their natural parchment, are among the most unusual of tomato novelties.

in water. The tomatillo is sometimes called the Mexican ground cherry in garden catalogs.

3) The Peruvian Cherry or Cape Gooseberry (*Physalis peruviana*) is a tender perennial that needs protection during the winter. It will grow year to year when temperatures are not persistently below 45° F., but otherwise should be treated as an annual like other ground cherries above. Growing one to three feet high, the plant produces a deep-gold-husked fruit that resembles a cherry in size and shape. The husk easily breaks to release its fruit.

4) The Golden Berry, an interrelated cross of forms of *Physalis,* is the very newest form of ground cherry and has just this year reached the American market. Cultivated like *Physalis pruinosa* above, the Golden Berry yields an average of four pounds of fruit per plant. The delicious juicy-sweet fruits, a dessert, are in great demand at top London restaurants. Golden Berry seeds are only available from England's Thompson & Morgan Seedsmen (see Appendix IV) who introduced them in 1976.

In a dry condition and in their husks, all ground cherry fruits will keep for months. Rich in vitamins, delicious and novel, they make a worthy addition to any tomato garden. As pointed out, their propagation and culture is very similar to that of ordinary tomatoes. There are differences, however—especially in seed propagation—so follow seed packet directions carefully. Husk tomato seeds are available from Burgess Seed Company, Farmer Seed & Nursery Company, Jung Seed Company, Olds Seed Company, and Gurney Seed & Nursery Company. (See Appendix IV for addresses of these firms.)

SPACE-SAVING POTATO-TOMATOES

Here's a novelty I've tried, *but can't recommend unless you're willing to take the risk.* Nevertheless, tomatoes and potatoes can be grown on the same plant, their roots intertwined. Just raise your tomato seedlings as you normally would (see Chapter Six) and then transplant the seedlings into one-inch holes filled with soil that you have made in whole seed potatoes (sprouting potatoes or potatoes with "eyes" will do). Lay the potatoes in a shallow flat filled with soil and wait until the tomato roots grow through the potatoes and into the soil. When transplanting the tomatoes to the garden, set the plants (potatoes and all) in one-foot deep holes in rich garden soil. This novelty is certainly a space-saver, yielding tomatoes on the plant above ground and potatoes below. (The potato plant will also send up vines, of course.) The chief drawback is that potatoes can transmit several diseases to tomatoes, and vice versa, which is the reason potato fields on farms are always widely separated from tomato fields. I had no such trouble, though, perhaps luckily. If you have an out-of-the-way area available you may want to experiment. Many such experiments have been made with tomato plants. At North Carolina State University, for example, tomatoes were grafted on tobacco plant roots. The result was a tomato with a high nicotine content! One of the earliest experiments with tomatoes, in 1910, produced a tomato-eggplant chimera having characteristics of both parents on the same branch. Lakeland Nurseries (see Appendix IV) offers a potato-tomato, or pomato, ready to plant if you don't want to go to the trouble of preparing your own.

The Pomato, a cross between potatoes and tomatoes, can be bought in this form or made at home.

UNUSUAL TOMATOES FOR PROBLEM SPOTS AND SPECIAL PURPOSES

There are tomato varieties for almost every problem spot or special purpose. The following are just a few of the more interesting ones. Check for other characteristics of these plants under the early- and main-season master lists (Appendixes I and II):

Border-plant Tomatoes—Tiny Tim, Yellow Tiny Tim (both only a foot or so high), and other midget varieties serve well as border plants in the flower or vegetable garden.

Canning Tomatoes—See Chapter Seventeen for many canning varieties.

Direct-seeding Tomatoes—Fireball, Foremost E-21, New Yorker, Coldset, Sub-arctic, and other early varieties can be seeded directly into the garden (see Chapter Four) instead of being started indoors.

Disease-resistant Tomatoes—See Chapters Twelve and Thirteen.

Drought-resistant Tomatoes—Porter, Red Cloud, Hotset, Summerset, Pink Deal.

Indoor and Greenhouse Tomatoes—See Chapter Fourteen for numerous plants to grow indoors.

Rot-resistant Tomatoes—The Greek variety Thessaloniki, bred especially for its storing qualities, keeps for a long time after picking even when fully ripe.

Sandy-soil Tomato—Starfire performs well in sandy soil.

Seedless Tomatoes—Contrary to some reports, there is no seedless tomato variety yet, but plant breeders say that seedless varieties (a dubious contribution, anyway) will be developed in less than a decade. Seedless tomatoes *can* be grown by using hormone sprays. (See Chapter Seven.)

Stakeless Tomatoes—See Chapter Nine for many plants needing no staking.

Stuffing Tomatoes—Burgess Stuffing, Ruffled, and Liberty Bell.

Sweet Wine Tomato—Giant Belgium is used to make old-fashioned tomato wine. (See recipe, Chapter Eighteen.)

Roma VF, long a top canning tomato, is also good for eating out of hand.

Unique Tomato Plants to Grow for Profit—Pin money, at the least, can be made growing many unusual tomato plants from seed if you have a greenhouse or enough sunny window space. The tomato plants can be sold direct to customers or wholesale to a nursery. Not many nurseries offer more than the most popular varieties—they can't afford to. Potted plants of many varieties mentioned in this chapter won't have much competition anywhere.

Wet Area Plant—Pearl Harbor is especially good for low damp grounds.

THE BEST
REGIONAL
TOMATOES

If you've had trouble growing tomatoes in your area, the United States Department of Agriculture particularly recommends the following varieties "to be grown in the home garden for taste and disease resistance" in different sections of the country. Most are main crop varieties. Check Appendixes I or II for particulars and other varieties developed for specific regions:

NORTH: Fireball; Morton Hybrid

EAST: Campbells 17; Fireball; Heinz 1350; Morton Hybrid; Spring Giant; Supersonic

Fantastic, a prolific plant especially good for the Midwest and East.

Cal-Ace, one of the best plants for western areas.

Homestead 24, a good southern variety.

WEST: Ace; Marion; Moscow; VFN-8
SOUTH: Atkinson; Floradel; Homestead 24; Marion; Nemared; Pinkdeal; Supermarket; Tropic-Gro
SOUTHWEST: Ace; California 145; Calmart; Early Park ✕7; Pearson; Plainsman; Porter; Red Cloud; VF13L; VF36

Some tomatoes do well almost everywhere. Keep your eyes open for All-America winners advertised in the nursery catalogs when selecting tomato seed. These are varieties tested in field trials across the country by the All-America selection committee of the American Nurserymen's Association and awarded gold, silver, and bronze medals as excellent plants for all America to grow. Burpee's Globe, Burpee's Jubilee, Floramerica, Pritchard, Small Fry, and Spring Giant Hybrid are several winners.

ALL-AMERICA
TOMATOES

Floramerica, an All-America prizewinner that does well just about anywhere.

Burpee's Big Boy, one of America's favorite main-season tomatoes.

Early tomatoes bear fruit in forty-five to sixty-five days from the time plants are set outside. They are excellent in quality, though usually smaller and not quite as tasty on the whole as main crop varieties. There are far more determinate or semideterminate (bushy) plants, which do not require staking, among the early varieties. Early varieties are also the most suitable for sections of the country with short growing seasons. A list of over one hundred early varieties, giving each plant's specific characteristics, can be found in Appendix I.

THE EARLIEST
TOMATOES

Main-season tomatoes, by far the most numerous of all varieties, bear in sixty-six to eighty days, while late-crop varieties yield fruit in about eighty-one to one hundred days. Both are usually large hybrid, indeterminate (tall growing) plants that must be staked and can be pruned without harming the plants. They bear the largest fruits, especially the late-crop varieties, and are usually the tastiest tomatoes. Difficult to grow in sections of the country with short growing seasons, they are a must everywhere else. See Appendix II for a detailed list of some three hundred main-season and late-crop tomatoes.

MAIN-SEASON AND
LATE-CROP
TOMATOES

See Appendix III for complete instructions by a prominent horticulturist on how to develop new unique varieties suited exactly to your own taste and unlike any other tomatoes anywhere.

DEVELOPING A
TOMATO VARIETY
JUST FOR YOU

Complete characteristics of every tomato mentioned in this chapter, including any special disease resistance can be found in Appendixes I or II. You'll also find therein hundreds more delicious tomato varieties to choose from.

Chapter Three

FROM THE SAHARA TO SIBERIA:
SUNLIGHT AND SOIL NEEDS FOR
TOMATOES EVERYWHERE

The adaptable tomato will grow outdoors practically everywhere except on the banks of the River Styx if the right varieties are planted and they are properly cultivated. Avid gardeners harvest tomatoes in the gelid northland, despite short growing seasons . . . in the sweltering heat of the tropics . . . in places where it rains close to half of the year . . . and in virtual desert areas where there may be only a few weeks of precipitation. Tomato plants will thrive in many types of soil, too. They are raised successfully all over America: on clayey soils in the Northeast; on highly alkaline western soils; on coral soils in Florida; even in the pure sand of beach areas, if they are fed daily with a water-soluble fertilizer. Nevertheless, there are certain ideal requirements that must be met to insure abundant tomato crops year after year. These should be pondered long before any seeds are ordered. Following is a review of the best tomato growing conditions and hints on how to achieve them—or how to make the best of what you have. . . .

TOMATO SUNLIGHT
AND SHADE NEEDS

For best results, tomatoes should occupy the southernmost exposure in the garden, where they receive about eight hours of sunlight on a bright day. Planted there they will grow, taste, and look better, be subject to fewer diseases and insect predators, and yield far more fruit than if planted in the shade. Tomatoes raised in full sun will even boast a higher vitamin C content. The USDA reports that Russian researchers protected tomatoes from sunlight in paper sacks and that these fruits had only one fifth the vitamin C content of tomatoes raised in full light.

Some gardening manuals put the bare minimum of direct sunlight for tomatoes at six hours a day, preferably in the middle part of the day. The tomato is definitely what is called a long-day plant, but this is not to say that tomatoes can't be grown and grown fairly well in much less direct sunlight. What's more, a bit of shade is good for tomato plants. Commercial growers often plant their tomatoes only fourteen inches apart so the blossoms get a little shade and fewer flowers drop off during hot spells—blossom drop being common when day temperatures rise above 90° F. and night temperatures go over 85° F. Shade also keeps tomato roots cool and helps protect developing fruits from sunscald; contrary to popular belief, intense direct sunlight doesn't redden tomatoes but burns them yellow. Shade can even help prevent uneven coloring of tomato fruits, as the red pigment in tomatoes refuses to form when temperatures are above 86° F.

It is important to remember that shade in one area of the country is not always the same as that in another. In a climate with a full quota of clear sunny days, four hours of sun is something different than four hours of sun in an area like the New England coast, where fog and cloud cover throughout much of the growing season results in about 65 per cent of available sunshine. Local conditions must be taken into account when planting tomatoes in the shade and, usually, experience is the only good teacher.

Tomatoes grown in partial shade *will* yield fruit, but less fruit in more time. Many gardeners forced to plant in half-sun have raised tomato plants yielding up to forty fruits on a bush. For years I have set several tomato plants in an area where they receive only three hours of direct sunlight a day. With good care the plants yield an average of about eight large-size fruits each. The first ripe fruit—not quite as tasty as tomatoes grown in full sun—is taken from these Burpee Big Boy plants about ninety days after setting the plants out, which is roughly two weeks late for the variety. We often pick ripe fruit through October after that, harvesting green tomatoes up until Thanksgiving in years when killing frosts come late. The plants tend to be tall (up to twelve feet) and thin-stemmed in reaching for the light. Their yield is small compared with plants set in the sun, which routinely bear twelve to twenty fruits and can yield up to two hundred tomatoes.

In short, it pays to plant tomatoes in full sun and to get as much sun as possible on the tomato patch by trimming overhead branches or even pollarding trees (cutting the branches back to the trunk). Certainly don't give up if your property is mostly shady. Experiment. Try different varieties, especially early-bearing types, from year to year. But as a general rule, plant your tomatoes in as much sunlight as you have. Plant under the branches of trees or in the shadows of tall buildings only if absolutely necessary, and be sure to calculate sunlight in the garden by observing the site at different times of the day and season. A spot that seems sunny in the morning may not be sunny in the afternoon due to shadows from trees or buildings. And as the sun drops lower in the sky with the progression of summer, obstructions to the east, south, and west will cast longer shadows.

TOMATO
SOIL NEEDS

The tomato isn't a finicky plant and will grow well in any reasonably good garden soil. In fact, soil that is too rich in nitrogenous matter isn't ideal for tomatoes, encouraging green growth at the expense of fruiting.

Most "tomatologists" claim that the ideal soil for early tomato varieties is a sandy loam; that is, a soil which contains much sand, but which has enough silt and clay to make it hold together. Squeezed when moist, sandy loam will form a ball that bears careful handling without breaking. Sandy loam is doubtless the best soil for all-around garden purposes and is ideal where earliness is the prime consideration. It can be worked almost as early in the season as sandy soil and is much more retentive of moisture and nutrients than are the sands.

For midseason and late tomatoes, a loamy soil is usually preferred. Loamy soil is even more retentive of moisture and nutrients than sandy loam and if squeezed in the hand when *dry* will form a ball. Because loamy soils hold moisture better than sandy soils there is less likely to be blossom-end rot on fruits planted in them—blossom-end rot being partly caused by fluctuation in water supply (dry weather one week and wet the next).

There are two ways to test soil fertility for tomatoes. You can either set the plants out and see how well they do the first year (unless the soil is very poor indeed on your property, you'll get a yield ranging anywhere from fair to excellent, or you can take soil samples and have them tested by your state agricultural extension service, or a commercial laboratory. There are also inexpensive do-it-yourself soil-testing kits widely available at garden centers. As one extension specialist puts it: "Gardening without a soil test is like building a house without a blueprint. It can be done, but it's harder and far more time-consuming in the long run." A good private soil-testing service giving prompt service and suggesting corrective measures is S. R. Sorensen, Prescription Soil Analysis, Box 80631, Lincoln, Nebraska 68501.

SOIL ACIDITY AND
ALKALINITY FOR
TOMATOES

While you're having the tomato-patch soil tested for fertility, have its pH tested, too (or do so with a soil-testing kit). The term pH is used to express the acidity or alkalinity (so-called "sourness" or "sweetness") of a soil. The symbol pH just means "potential hydrogen," and indicates the breakdown of a soil solution into positive hydrogen ions and negative hydroxyl ions. In an acid soil hydrogen ions would predominate, while in an alkaline soil there would be more hydroxyl ions. A soil's pH is expressed as a number ranging from pHO to pH14. Any reading below pH7 is acid, pH7 is neutral, and all values over pH7 are alkaline. The tomato is a moderately acid-sensitive vegetable that can be grown successfully in a pH range of from pH5.5 to pH6.8. It should be noted, however, that tomatoes grown in an acid soil just do not taste as good as those grown at a neutral pH of about 6.8. At any rate, pH test results will include instructions on how to correct your soil if it is too acid or alkaline. Lime (agricultural-grade ground limestone) is usually added if a soil is too acid, aluminum sulfate (or organic materials like sawdust, oak leaves, and cottonseed meal) if the soil is too alkaline. Specific application rates will be given by the testing service.

There is no way to accurately tell the exact pH of a soil without a soil test. So don't lime the tomato patch as a matter of course every year—you might be causing trouble at worst and losing money at best. The only rule of thumb that can be given here, besides testing, is that most soil in America is slightly acid, so chances are that your garden soil is in the safe range for tomatoes. It's also true that most soils that will grow good-sized carrots or beets do not need lime, and that if the lawn or flowers on your property are thriving, you should leave well enough alone.

Fortunately, almost anything wrong with garden soil can be corrected if heavily supplied with organic matter. By all means follow any specific recommendations for improvement that are made as a result of soil tests; but if you don't test your soil, you'll find that a garden laced with organic matter from year to year will stabilize at a good tomato pH of about 6.8 and be water-retentive and rich in most nutrients as well. Make it a point to maintain the garden this way whether you choose to test the soil or not. Here are several simple methods:

LONG-TERM SOIL IMPROVEMENT IN THE TOMATO PATCH

1) Green Manure Cover Crops—thickly sow the garden with either red clover, vetch, annual grass seed, rye, buckwheat, oats, or millet toward the end of the growing season. Fertilize the cover crop when it comes up and dig it under the next spring. Legumes like red clover and vetch are most valuable, adding the largest quantities of nitrogen, phosphorus, and potassium to the soil, but all of the above "green manure" crops are good soil conditioners. Their roots will also loosen heavy soils and help hold the particles of sandy soils together. Write for the free booklet *How to Do Wonders with Green Manure,* from Garden Way Manufacturing, 102nd Street and North Avenue, Troy, New York 12180.

2) Organic Materials—at the end of the growing season dig organic materials such as decayed leaves, hay, grass clippings, and manure about one foot into the soil. Whole leaves that aren't decayed can also be used, although they are harder to work with and take longer to break down. One way to remedy this is to put underaged leaves through a shredder first, or run through them a few times with a power mower.

A more thorough way to improve soil organically is to cover the area to be dug or plowed with five inches of vegetative material (decayed leaves, peat moss, ground corncobs, chopped straw, grass clippings, compost, or what have you) and sprinkle on applications of fifty pounds each of superphosphate and muriate of potash for each one thousand square feet. Dig or plow this under to a depth of about one foot.

Most organic materials will decay and become part of your new soil by spring planting time. All will certainly break down by the end of the following summer. Improve soil like this for two or three years and you'll have a highly satisfactory tomato patch no matter how poor the soil was to begin with.

3) Mulching—when the soil warms up in early summer, mulch the garden with any organic materials you have on hand. Mulching (see Chapter Eleven) simply means to cover all the soil with six inches or more of organic or vegetative matter. The organic matter will feed plants, preserve moisture, and break down to enrich the soil. Over the years a more fertile soil will be created.

There are many books available devoted exclusively to methods of soil improvement, and you should certainly consult one if your garden badly needs renovation. But keep in mind that very few soils indeed won't grow some tomatoes while you're busy improving them, especially if the plants are supplemented with additional fertilizer.

LAST-MINUTE TOMATO-PATCH IMPROVEMENT

This short section is aimed at the prospective tomato grower who has decided to grow tomatoes this spring and has a soil so poor that it seems impossible. It isn't—tomatoes can even be grown in beach sand. Some quick solutions follow:

▪ Circumvent the problem temporarily by growing tomato plants in containers (see Chapter Five) while you're improving the soil for next year.

▪ Make nutrient-fortified planting holes for your tomato plants. Simply dig a 2-by-2-foot hole for each plant and fill it with a mixture of one-third topsoil, one-third compost or peat moss, and one-third builder's sand. All of these ingredients can be purchased at any garden store.

▪ Dig holes for your plants and fill them with Tomato Soil, a new product designed specifically for tomatoes that also contains all the necessary nutrients for the growing season and doesn't have to be fertilized.

■ Buy a truckload of good topsoil from a garden supplier or farmer and spread it over the ground at a height of about six inches, framing the area with boards or concrete blocks and creating an instant raised bed. A good alternative is to fill most of the raised bed with leaves or other organic matter, sprinkle on rooted manure or fertilizer, and cover the pile with soil. Or you can just pile large mounds of good topsoil over your poor soil and plant a tomato in each mound—in this case, however, be sure to leave a slight depression in the top of each mound to collect water.

■ On very clayey soils (soils so stiff that air won't penetrate) dig vertical drains near your tomato plants and fill them with gravel. This will facilitate watering and get oxygen to the plant roots for growth.

■ To grow tomatoes in pure beach sand—say in dune areas near a summer cottage—strip off the sand to a depth of about a foot. Then lay plastic over the bottom of the ditch and spread several inches of shredded leaves or other organic matter over the plastic. Refill the trench with the sand removed, adding potash and phosphorus to it at the rate suggested on the bags. Before the plastic rots out it will serve as a water barrier, preventing water and nutrients from leaching away. By the time the plastic does disintegrate, rotting roots and other vegetation will have formed a true soil in the treated area. Tomatoes planted in such soil, even the first year, won't have to be fertilized any more than plants in heavier soils. The only other way to grow tomatoes in sand is by constant daily watering with a weak fertilizer solution, which won't yield results nearly as good.

DRAINAGE FOR TOMATO PLANTS

Good drainage is a must for tomatoes, which dislike "wet feet." Never set out tomato plants where puddles form after rains or waterings—the plants won't grow well in such spots and poor drainage is associated with a number of tomato diseases, including bacterial wilt. Good drainage means that water passes *through* the soil, not over it, and that water doesn't collect on the surface. Good drainage is necessary because when water replaces the air in soil, roots suffocate—the roots can't develop properly without a constant supply of oxygen and removal of carbon dioxide. A soil that is well drained is one in which water moves through quickly, never completely blocking the movement of air through the soil.

If you're not sure whether your soil is properly drained, just make a simple test. Dig a 2-by-2-foot hole in the garden area in question and fill it to the top with water. If the water is gone in from one to two hours, drainage is all right. Soil that takes over an hour to drain is too heavy or clayey. Soil that drains in much under an hour is too light or sandy.

Most problems with poor drainage come from heavy clay soils which drain too slowly. Definitely refrain from planting in any soil unless it forms a ball when molded with the hands and crumbles easily when pressed with the fingers; it is too wet to handle if you can't do this with it. Try to improve heavy clay soils by digging in compost and other organic matter every year. Mulching in warm weather will also help.

Sandy soil that drains too quickly to supply plants with full nourishment is also a problem in growing tomatoes, though not generally as serious. Such light soil has to be supplemented with water when it doesn't rain for five to ten days and fertilized more often than a heavier soil. In this way tomatoes can easily be grown on it. However, try to improve the drainage of sandy soils each year, too. Follow the same program as you would for clay soil.

Another problem related to good drainage is runoff. If water cascades down a slope so heavily that gullies are formed, the slope is too steep for planting. You can try to terrace the slope, running the tomato rows squarely across it to slow down water runoff, but if you have any choice, plant in a different area.

Really serious drainage problems can be permanently corrected in several ways. One is by laying underground tiles at a depth of about two feet. The special tiles, installed on a slight slant and aimed at a low-lying area, will direct ground water seeping up into them away from the planting site. Be sure, however, that you have a low-lying area for an outlet, or you'll accomplish nothing but the creation of a sump into which runoff from the ground will settle. Don't dig open ditches for drainage either—they're unsightly, breed mosquitoes, and are rarely effective.

Putting in a good underground-drain tile system can be an engineering job, is always arduous work, and may not be wholly acceptable even if done perfectly. The best alternative is to grow plants in containers where they can't be placed directly into the garden soil (see Chapter Fourteen) or to make a raised bed as described in the previous section on soil improvement. Don't forget, too, that many tomato varieties that do better in dry, damp, heavy, or sandy soils are noted in the extensive variety lists in these pages.

Chapter Four

THE EASY WAY:
FOUR TOMATO-GROWING SHORTCUTS

Growing healthy tomato transplants from seed is a relatively simple process, doesn't require any backbreaking work, and saves money in the bargain. More important, it enables you to choose from a greater number of varieties and have a quality control on your plants from the very beginning to the end. However, since many people prefer to buy plants in various stages of growth from nurserymen— or to seed tomatoes directly into the garden—let's examine these shortcuts before turning to growing transplants from seed indoors.

Potted tomato plants should be purchased just before planting time, not kept in the house for several weeks. A few rules for selecting these full-grown transplants follow:

BUYING FULL-GROWN, POTTED TRANSPLANTS

- Purchase named varieties from a reliable nursery.
- Buy stocky, thick-stemmed transplants of medium height.
- Don't buy thin or leggy plants, no matter how tall they are.
- Be sure all transplants have six to eight true leaves.
- Foliage should be a deep green color.
- Transplants should be potted individually in at least four-inch pots, not crowded together in flats with their root systems intertwined.
- Avoid any plant with the slightest trace of a disease or nutritional deficiency —yellowish, reddish, or light green leaves.
- Be certain all plants are insect free.
- Don't buy old, hardened transplants. Plants grown ten to twelve weeks or more (excepting those in very large containers) are usually stunted in growth, their fruit small and their yield slim.

▪ Never buy a plant with flowers or fruit on it. One of the most common mistakes gardeners make is to purchase expensive transplants already setting fruit with the idea in mind that they will have earlier tomatoes than anyone in the neighborhood. While they do yield a few early tomatoes, these costly plants do so at the expense of their vegetative growth. As a result, plant growth and total yield are drastically reduced. I have time and again seen young, healthy transplants outgrow such giants within a few weeks and go on to produce far more fruit. In fact, it is best to remove flowers and fruits from all transplants—this will enable the plants to establish better root system and produce more foliage for photosynthesis.

▪ Plant transplants in the garden just as you would transplants grown indoors from seed. (See Chapter Seven.)

Often you can purchase prestarted tomato plants grown from seed in peat pots. Easy to transplant, these are set out in the garden, pot and all, at planting time.

Bare-root tomato plants (unpotted) are not generally a very good idea, but they are cheaper than potted transplants and some people buy them direct from large nurseries, where they are pulled out of the ground and wrapped in damp newspaper, or order them airmail from the few seedsmen that handle them. As with potted plants, be sure you are buying disease-free, healthy plants in named varieties. It is even more important to get bare-root plants in the ground outside immediately after you receive them. If you don't, they'll soon die. Shade them from the sun until they recover from the shock of transplanting, feed them a weak solution of liquid fertilizer, and keep the soil wet around them at least for the first week. Large bare-root plants that look almost dead when put in the ground often withstand transplanting shock if they are treated this way, and go on to produce fifteen or more fruits. But the method is risky, nothing is to be gained over other alternatives, and the plants frequently die. Decidedly not recommended except where the plants are extremely cheap or you want to experiment. Burgess Seeds offers a fairly large selection of hardened field-grown bare-root plants at twelve for $3.95, or fifty for $10. Earl May Seed & Nursery Company also offers a good number at about the same prices.

Choose seedling flats in nurseries with as much care as larger tomato transplants. Named variety, disease-free seedlings are what you need; don't take anything less. Seedlings that look thin and leggy, or are crowded together and unwatered, indicate poor growing practices—look elsewhere. As for care of these little seedlings, they should be treated in the same way as plants that you raise from seed. (See Chapter Six.)

Probably 95 per cent of all gardeners plant their tomatoes from potted plants or transplants grown indoors. Few people realize that early (and even a few main season varieties) can be seeded directly into the garden throughout most of the United States. Only in extreme northern areas with short growing seasons is this impossible. In fact, many commercial growers don't bother at all with transplants and seed directly into the ground. The major disadvantage is that very few varieties can be seeded directly.

Direct-seeded plants are generally healthier in most respects, usually bear well, and varieties like Coldset often bear earlier. The seed lies dormant until suitable weather arrives, by which time all danger from aphids carrying many tomato diseases is over—resulting in healthier plants. If the weather is good, direct-seeded plants will bear even before main crop transplants and if it isn't, they will bear just as soon.

The main secret in direct sowing of tomatoes is to choose seed of early varieties. New Yorker, Fireball, Coldset, and Subarctic are particularly recommended, as are cherry types like Small Fry or Tiny Tim. For a main season variety, with larger fruit, try Foremost E21 (seventy days). Check the variety lists in Appendixes I and II for further characteristics of these plants and other possibilities. But remember never to direct-seed a variety that takes much longer than seventy days to yield fruit and to choose stakeless bushy determinate types whenever possible.

Warm up the soil for direct seeding as you would for any early transplant. (See Chapter Seven.) Plant on a well-drained southern slope, avoiding frost pockets, about a month before you would set out main crop transplants and when all danger of frost has passed (May 10 in the North, March 20 in the South, April 15 in the Midwest). A good time to plant, if you observe natural signs, is when the flowers first appear on maples. First fertilize and lime the soil as described in Chapter Eight. Then spade to a depth of about one foot and rake the seedbed smooth. Straight rows can be easily made by stretching out a rope or garden hose and walking on it—this will leave a slight indentation in the ground that you can use as a guideline. Plant the seed about a quarter inch deep (deeper than usual) and one to two inches apart. Water seeds and seedlings with water that is at least of air temperature.

Begin to protect the seedlings as soon as they emerge from the ground, using one of the many protection methods given at the end of Chapter Seven. When the plants are three inches high, thin them so that they are 1½–2 feet apart in the row—except for the cherry types, which can be grown at about nine inches apart.

When you first try direct seeding of tomatoes, plant only a short row of them. Put out transplants in another row or two as added insurance.

TOMATOES FROM VOLUNTEER SEEDLINGS

Volunteer seedlings—plants that sprout from fruit discarded in the garden the year before—should not be removed from their growing sites if you want to keep them. If they are, their long tap roots will be broken. Otherwise treat them exactly as you would direct-seeded plants. But remember that seeds from hybrid tomatoes will not produce plants true to variety. (See Chapter Sixteen.)

HOW MANY PLANTS YOU'LL NEED

In most cases a tomato plant will bear from twelve to twenty fruits, or eight to ten pounds of fruit, but there have been a number of cases on record of plants, always rampant-growing main-croppers, bearing over two hundred fruits. Tomatoes easily yield eighty portions per ten feet of row. Generally speaking, four plants per person would be sufficient for a family, but how many plants you need depends on how sunny your yard is (tomatoes yield less when out of full sun), whether you intend to do any canning, and, of course, how much you like tomatoes. If you want to can tomatoes for a year's supply, for example, figure on ten plants per person.

A final suggestion: If you have the space, make an insurance planting of tomatoes, especially when setting out early varieties. Wise gardeners do this with all crops, as can be seen by the old seed planting jingle: "One for the rook,/one for the crow,/one for the starlings/and one to grow." And if *all* grow, well then there will be a lot of tomatoes to give away, or make tomato wine with.

Chapter Five

CONTAINING 'EM:
SOILS AND CONTAINERS FOR GROWING
TOMATO SEEDLINGS INDOORS

If you decide to start your tomatoes from seed indoors instead of buying full-grown plants, carefully select the containers and soils to be used before planting the seed. This will save much time and trouble in the long run and help guarantee healthy, vigorous seedlings. Many containers and soils are discussed in this chapter.

Tomatoes will germinate in almost any soil. But if soil from the garden—the most inexpensive medium—is used to grow tomato seedlings, the lighter it is, the better. In fact, soils rich in nutrients are usually rich in disease-causing fungi and bacteria as well, creating more problems than they solve. If you don't have a light, sandy soil in your yard, you can make it by mixing together equal parts of coarse builder's sand, peat moss (or leaf mold), and garden loam (topsoil). Other good mixtures for seedlings are one part peat moss, two parts sand; and one part compost, one part sand, and two parts loam. Any garden soil or mixture used, however, should always be sterilized to cut down on the chances of diseases.

STERILIZING
TOMATO SOIL

Despite their low cost, garden soil or soil mixtures aren't the best mediums for starting tomato seedlings—for they contain weed seed and various disease-carrying organisms. Especially dangerous is damping-off disease, which is caused by several fungi in soil. Damping-off disease can be identified by the fuzzy white coating it leaves on damp soils. It either causes seed to decay before it sprouts, or kills off young seedlings. Protect against weed seeds, damping-off, and other harmful fungi, bacteria, and insects by sterilizing tomato soil from the garden in one of the following manners:

Baked Soils. Place the soil to be used for planting tomato seedlings in a shallow pan (not deep coffee cans or the like) and bake in the oven at 180° F. for forty-five minutes. All the soil must reach 180° F. before the job is done. A good way to tell is to put a large potato in the oven along with the soil. When the potato is baked, the soil is sterilized. Be sure not to overcook, as overcooking can release toxic substances in the soil.

Pressure-cooked Soils. Cook the seedling soil in a pressure cooker at five pounds pressure for twenty minutes.

Formaldehyde Treatment. To a cup of water add 2½ tablespoons of formaldehyde or formalin. Sprinkle the mixture on a bushel of soil, stirring it well. Wrap the basket in a blanket three days or until the smell is gone before using the soil.

Disinfectants. Most nurseries sell disinfectants that seeds are easily coated with to help prevent damping-off disease. Instructions come with the packages.

Covering unsterilized soils with a thin layer of a sterile medium like sand, after you've sown your tomato seed, will also help control diseases somewhat. Or you can use potting soil, which is relatively sterile and is sold in most garden outlets. Still another solution is to use fine Canadian or German horticultural-type sphagnum peat moss as a partial filler in tomato seed flats. Fill the flat with unsterilized garden soil up to about one inch from the top and place half an inch of sphagnum moss on the surface of the soil. Then sow the seed and cover it with more sphagnum moss that has been sifted through a fine screen. Don't use sphagnum moss alone—it doesn't have enough body to hold up large seedlings.

Sterilizing garden soil or soil mixes, no matter how carefully done, does not result in *complete* sterilization. Safer for seedlings, though more expensive, are various naturally sterile artificial soils that are completely free of weed seeds and disease-producing organisms. Following is a list of them giving their advantages and disadvantages:

SAFE ARTIFICIAL
TOMATO-SOIL
MIXTURES

Vermiculite. Vermiculite is actually a lightweight, expanded mineral—mica that has been heated at 1,800° F. until it breaks into tiny fluffy pieces. A completely sterile substance containing magnesium and potassium, vermiculite holds and releases large quantities of both minerals and water for plant growth. Another advantage in using vermiculite is the ease with which tomato seedlings can be lifted from it without any damage to their roots. However, like sphagnum peat moss, vermiculite doesn't have the body to hold up large seedlings—in other words, it's perfect to start seeds in, but not satisfactory if you intend growing plants in the mixture until they're set out in the garden. It is also devoid of nutrients, save for minerals, and seedlings grown in it have to be fed weekly with a liquid plant food after they reach about 1½ inches in height, or must be transplanted to a more nutritious soil mixture. Use only horticultural grades if you use vermiculite, not the insulation grades.

Perlite. Perlite is a form of volcanic rock that has expanded after being heated at very high temperatures. Sterile like vermiculite, it is slightly heavier, and contains no mineral nutrients. Vermiculite is generally preferred for starting seeds.

Blended Synthetic Soil Mixtures. These are the latest and best development for starting tomato seeds and are highly recommended. Blended synthetic soil mixtures are composed of peat moss, vermiculite, and various fertilizers. All you have to do is toss a bag or so of one of these special mixtures in a pail, add water, and stir until the mix is damp (not sopping wet). Not only is this medium lightweight and free of disease, insects, and weed seeds, it is also dense enough to hold large seedlings and is supplemented with plant nutrients. Seedlings can easily be lifted out of flats containing these synthetic soil mixes for transplanting into individual pots. What's more, the same kind of mixture can be used for the repotting job. Some good brands are Jiffy Mix, Kys-Mix, Pro-Mix, and Redi-Earth. Other mediums are only better in that they are cheaper.

Homemade Synthetic Soil Mixes. Money can be saved by making a bushel of your own synthetic soil mix, which will be just as good as any you can buy. Just thoroughly mix in a large plastic bag: half a bushel of Canadian or German sphagnum peat moss; half a bushel of #2 size horticultural vermiculite; eight tablespoons of 6-12-6 (or 5-10-5) fertilizer; and five tablespoons of powdered superphosphate. Use clean garden tools when mixing so that the sterile mix will not be contaminated with disease organisms in the soil or debris. (Wash the tools with a disinfectant if necessary.) The fertilizer in this mix, which was developed and tested at Cornell University, is sufficient to support plant growth for four to five weeks. Before using a portion in seed flats or pots, make sure that it is thoroughly wet. (A small amount of detergent can be added to the water if the mix proves difficult to wet.) Keep the medium moist until the seedlings are established. Then water as needed, using a 20-20-20 liquid fertilizer for the watering after a few weeks.

Easy-to-make containers to fill with your planting medium include everything from wooden seed flats and standard clay pots you might have on hand to pots made out of old milk containers and pots fashioned from detergent bottles. Their chief defect is that tomato seedling root systems are disturbed to some extent on transplanting out of them. On the other hand, homemade containers are free for the making or finding. Always provide drainage for these containers and cover the drainage holes with small stones, pieces of broken shard, or bits of charcoal to prevent dirt from clogging them. Some homemade containers are described below, but by no means limit yourself to these. Aluminum trays from frozen foods, coffee cans, tomato cans, cake pans, old frying pans—all are possibilities. Any container that can hold soil and hold its shape when wet will do:

Wooden Seed Flats. Any flat, shallow wooden box may be used, such as an old top drawer from a dresser or end table. Or you can construct your own seed flat, using inexpensive lumber scraps. The flats should be about three inches deep, fourteen inches wide and twenty-four inches long. Drainage can be effected by drilling holes in the bottom, or by lining the bottom with absorbent sphagnum moss before filling in the soil. Seedlings should only remain a few weeks in such flats before being transplanted to larger containers.

Plastic and Fiberboard Seed Flats. Just as good as wooden flats, these come in many sizes and anyone who has ever bought seedlings from a nursery has several lying around. Old plastic refrigerator trays serve just as well. Use the same as you would wooden flats, cutting and covering holes in the bottom for drainage if there aren't any.

Plastic Pots. The three-inch size is best for individual tomato plants. Bore holes in the bottom for drainage when necessary.

Clay Pots. The old-fashioned clay pot with large drainage hole remains an excellent growing container, but dries out fast and must be watered frequently. Use the three-inch size for individual plants and always cover the drainage hole.

Milk Containers. Cut a quart-size milk container down to about one third its size with a serrated knife. Then jab a few holes in the bottom for drainage, and partially cut the bottom so that it will come off easily when seedlings are ready to be transplanted. Milk containers are filled with soil and used either to germinate a number of seedlings for transplanting soon after, or for germinating single seeds that will grow in the container until transplanting to the garden. Half-gallon sizes serve just as well but are more bulky. Paper cups and styrofoam cups are used in the same way.

Margarine or Cottage Cheese Tubs. Simply fill these with soil after perforating the bottoms for drainage, and plant the seeds. Put the lids back on after making ten or so holes for ventilation so that moisture and heat will be sealed in for quick germination. As soon as your seedlings develop true leaves transplant them to larger containers.

Berry Boxes. Though soil spills out of these, making them a little messy, berry baskets provide excellent drainage for seedlings. If you line them with plastic to prevent soil leakage, punch drainage holes in the plastic.

Plastic Jugs. Especially good here are old Clorox bottles (thoroughly washed out). Cut one in half, fill the perforated bottom half with soil, and use for a seedbed. The top half can be fitted back on whenever protection or more heat is needed.

Egg Cartons. Fill with soil, plant a single tomato seed in each egg compartment, and close the top when protection is needed. Transplant the seedlings to larger containers when they develop their true leaves. Remember to punch holes in each egg compartment for drainage.

POTS YOU CAN
PLANT

So-called organic containers, available at all gardening outlets, are those which can be planted whole in the ground without removing young seedling plants and disturbing their roots. Drainage is no problem with them, they are often prefertilized, and many gardeners simply plant seeds in them and do little else save watering until they're ready to be planted in the garden. The only problem is that many experts contend that a tomato seedling should be transplanted to a larger pot at least once before being set outside. But then many of these containers can be transplanted into larger pots or used as larger containers themselves. Several are described below:

Peat Pots. Made of wood pulp and ground peat moss, this type of pot is filled with soil before seeds are planted in it. Often peat pots are sold in tray forms called peat strips—the trays containing individual peat pots that can be pulled apart and transplanted. Plants in peat pots are not supposed to suffer any checkback when transplanted into the garden, but in practice they often remain drier than the soil surrounding them, the pot taking quite some time before it breaks down in the soil. If you plant seeds in small peat pots, you can transplant them to larger ones, but such would be a costly practice.

A Jiffy-9 Peat Pellet ready to be transplanted outside in the garden.

Jiffy-Pots, peat pots made with soluble fertilizers, are among the most popular containers for starting seeds, but must be filled with soil.

Whenever peat pots are transplanted anywhere, be careful to bury the entire pot. Any part of one sticking out of the soil acts as a wick pulling moisture away from plant roots.

Jiffy 7 Pellets. The Jiffy 7 is a relatively new development—a compressed peat pellet containing fertilizer that expands to seven times its size when placed in water to make a 1-¾-by-2-inch container. The hard, thin chips, covered with

Vigorous plants in Big Red, the new easy-to-grow, tomato-shaped starter kit. The "tomato" container is easily broken in half, each half holding three seeded peat pellets that only have to be watered.

Jiffy-9 Peat Pellets below combine the functions of a pot and potting soil into one expandable shape-retaining growing unit. The compressed peat pellet makes for easy storage and the expanded pellet gives you a growing media with good air/water relationship, plus nutrients. The expanded Jiffy-9 (1½ inches in diameter and 1¾ inches high) can be used for starting seeds, seedlings, and cuttings. Also available in trays of twelve and in 2-inch-diameter size.

plastic net, swell up into large soft balls after soaking them for only a few minutes in a bucket of water. Then seed or a small seedling transplant can be set in the ball. The material is free from disease-causing organisms, insects, and weed soil. Each ball needs only to be watered and will nourish a single seedling and protect it from shock when transplanted either to a larger pot or into the garden. Jiffy 9 is a larger-size pellet.

BR8 Blocks. Similar to Jiffy 7 pellets, BR8 Blocks do not expand when they absorb water. These fiber blocks come in the shape of a two-inch square pot, contain fertilizer, and are also made of materials free from disease-causing organisms, insects, and weed seed. Simply water the block and plant with seed or seedling transplant.

Fertl-Cubes or Kys-Kubes. Made of a blend of mosses, plant food, and vermiculite, these tiny cubes have all the advantages of the above pellets. They are much smaller, though, and seedlings planted in them can easily be transplanted to a larger container when they make their true leaves.

Chapter Six

A SOWING PRIMER:
STARTING TOMATO SEEDS INDOORS

Once you select the soil and containers you prefer, it's time to start your tomato seed indoors. Tomatoes should not generally be planted *outside* in the garden until after the last average frost, when both air temperature and soil temperature are at least 60° F. (See Chapter Eight.) Therefore, in order to raise seedlings large enough to transplant outdoors at the proper time, seed must be sown *indoors* seven to eight weeks from the proper date in your locality for setting tomato plants in the garden. This is the rule in all save areas with long growing seasons like the South. You can easily determine local planting times by calling or writing your weather bureau or state agricultural extension service, or by checking with the local newspaper's garden columnist. It is best at first to follow the advice of these experts and experienced local gardeners as to planting dates, but within a few years you'll know from experience when is the best time for you to plant seeds indoors.

Only when gambling on planting seeds for early tomatoes (see Chapter Seven) should this planting date advice be ignored. Remember that the object is to grow healthy stocky plants, not spindly, root-bound ones. In fact, by starting seed a little *later* in spring (estimate your outdoor planting date as *two weeks after the safe date in your area*), your plants will have more daylight hours to grow sturdy in a sunny window. They'll also stand a better chance outside for another reason —aphids carrying serious virus diseases affecting tomatoes will have been wiped out by predatory insects when it finally comes time to transplant them.

PLANTING
THE SEEDS

After selecting the variety, soil mixture, and container you like best, you're ready to raise tomatoes from seed. All of the soil mixtures and containers previously described will work. In fact, I've often grown perfectly healthy tomato seedlings in unsterilized, sandy soil taken from the garden and placed directly in a cut-down milk container—doing nothing but putting in the seed, keeping the container in a sunny window, and watering until it was time to set the plants outside. Other people do no more than sow and grow their seedlings in pots containing established houseplants and seem to be successful every year. But what I believe is the best method for planting tomatoes from seed takes a little more time. It is described in ten steps below, though many of these pointers can be used when following any method:

1) Treat your tomato seed for disease-resistance if you've saved standard (non-hybrid) seed from last year or if the seed packet doesn't specify that this has been done. (See Chapter Sixteen.) Almost all commercial seed is treated, however.

2) Soak the seed in a mixture of water-soluble fertilizer for two hours before planting—recent tests prove that plants grown from fertilizer-soaked seeds are invariably heavier rooted and healthier.

3) Use a wooden seed flat for a container (see Chapter Five), lining the bottom for drainage, as explained, and filling it to within a half-inch of the rim with an artificial soil mix such as Jiffy Mix or your own artificial soil mix as described in Chapter Five.

4) Thoroughly water the soil mixture in the flat before seedling. (This can be done by submerging the flat in a sink of water.) Then firm the soil down with a flat board.

The easiest method, but not always the most successful, is to plant tomato seeds directly into a regular pot.

5) Estimate how many plants you'll need. (See Chapter Three.) Always plant twice as many seeds as this so that you'll have a good choice of plants when thinning later.

6) Do not plant seeds too deeply; if you do they may rot before they can germinate. A good rule to remember is to plant any seed three times as deep as the width of the seed. Set tomato seeds about a quarter inch deep and one inch apart. You can use the back of a knife, or holes can be punched with the eraser on a pencil, which is about a quarter inch long.

7) Cover the seed with planting mix and water gently with a fine spray or by a tin can with a dozen or so small nail holes punched in the bottom. If you water too heavily, you may wash out the seeds.

8) Seeds don't need light to sprout, but they do need moisture. Slip the moist seed flat into a "tent" made of a clear plastic bag. Tie it up loosely, sticking two pencils in the soil at each end of the flat to keep the plastic from touching the soil. This will help retain moisture so that the soil won't dry out. If the soil does appear dry, however, water it gently again. You can also use a pane of glass or a layer of newspaper to cover the seed flat. Whenever excess condensation (some is normal) appears on the plastic or glass, pull the covering back a little to allow the moisture to evaporate. Leave plastic in place until seedlings are ready to be transplanted to a larger pot.

9) Seeds also need warmth to germinate. A temperature of 65°–70° F. is good during their germination period. Keep them any place where that temperature is maintained until they germinate.

10) If you don't have a place where the temperature is a constant 70° F., the

Small plastic flat seeded with tomatoes and covered with plastic film to help maintain proper temperature and moisture for germination.

flat can be set over a radiator until the seeds germinate. But be certain the heat isn't too high. Temperatures over 70° F. may produce tall spindly sproutlings that will never grow thick and healthy. A more expensive way to provide bottom heat is by using cables, which can be purchased in any garden supply store. They are easy to install beneath the soil in seed flats and cost about $10 for thirty-six feet of cable.

THINNING-OUT SEEDLINGS

Your tomato seeds will germinate in seven to ten days. The first leaves to appear when the young plant breaks out from the soil into life aren't true leaves but cotyledons or seed leaves, which are filled with starch for the developing plant. As soon as the seedlings show their first true leaves (shaped like tomato leaves) and are 1–1½ inches high, they should be thinned. This gives the remaining plants more room to grow and greatly improves their quality. As much as you'll hate to do it, ruthlessly snip off every second plant (no matter how good it looks) at ground level with a pair of scissors. The only alternative is to carefully remove excess seedlings and transplant them to individual pots.

CHILLING THE SEEDLINGS

Here we have an important exception to the traditional 70° day and 65° night temperature requirements for growing tomato seedlings. The latest research reveals that tomato seedlings should be chilled as soon as the first true leaves appear—when the plants are 1–1½ inches tall. The seedlings are best chilled at 50°–55° F. for two weeks. Various studies made at Michigan State University show that chilled plants, which are chilled at the time flower number is being determined in the plants, bear earlier and heavier yields of fruit. Not only is flowering stimulated, but the plants become stockier and thicker stemmed, enabling them to survive transplanting better. The plants must, of course, have light while they are being chilled. After the plants have been chilled two weeks, go back to growing them at temperatures of 70° during the day and 65° at night.

SEEDLING LIGHT REQUIREMENTS

Light is probably the most important factor for good seedling growth. Tomatoes need about twelve hours of light a day, so it is important to keep them in the sunniest, southernmost window in the house, otherwise they'll become long and spindly. Light and temperature being equal, a bathroom window is preferable to a window in the usually dry air of a living room, as tomato seedlings like humidity.

AS AN ALTERNATIVE TO THE SEED-STARTING PROCEDURE DESCRIBED IN THE TEXT, YOU CAN USE THIS EASY INEXPENSIVE METHOD TO GROW TOMATOES FROM SEED.

Make shallow furrows with a stick or ruler.

Sow seeds as thinly as possible from packets.

Water newly planted seeds gently so as not to wash out seed. When young plants appear, be careful never to let them dry out.

When plants are about one inch high and have started to form true "tomato leaves," gently lift them from the seedbed.

Prepare another flat (or any container) for transplanting of young plants.

Set young transplants in the planting holes; firm them in place.

Thoroughly water the transplants immediately after planting. The plants can be transplanted again in three weeks or set out into the garden from this planter.

A kitchen window, no matter how warm and sunny, can be a fatal place to grow tomatoes if there is the slightest gas leak in the kitchen. The tomato is a very sensitive plant, particular about the air it breathes. Concentrations of gas too small to be detected chemically (1–1,000,000 parts of air) will infuse tomato leaves. Florists, in fact, use potted tomato plants as indicators to prove that gas is seeping into their greenhouses. The Missouri Air Conservation Commission recently suggested that tomatoes could be used to check pollution levels in the atmosphere, the sensitive plants serving as a warning device much as canaries have been used to detect poisonous gases in coal mines.

In any event, don't place tomato plants next to a cold window in *any* room. And remember to turn the plants every day. Otherwise they'll grow toward the sun—a phenomenon called heliotropism—and develop a permanent lean.

ARTIFICIAL
LIGHTING FOR
SEEDLINGS

If you haven't a warm sunny window in the house, your only choice is to make do with as much sun as you have (and risk weak, leggy plants), or use artificial lighting. Many types of gro-lamps and fixtures are available, ranging in price from ten to hundreds of dollars. Some come in stand forms and others are fixtures to be suspended from the ceiling. The wide-spectrum fluorescents introduced by Gro-Lux are particularly good. Two of these 40-watt tubes suspended a foot or so over a table will even supply a total light source for growing seedlings in a windowless basement. Indeed, Gro-Lux lights have been used to raise tomatoes in Alaska in the middle of winter. They can also be used to supplement window light where you have some but not enough sun. Directions come with the units.

You can make your own inexpensive artificial light by hanging two 48-inch, 40-watt fluorescent tubes about a foot above your seedlings. Supplement these with a couple of incandescent bulbs to provide the "red light" that all growing plants need. Keep the lights on eighteen hours a day, not twelve hours, when using any fluorescent lighting.

FERTILIZING AND
WATERING TIPS

▪ Where a prefertilized soil mixture is used, tomato seedlings shouldn't need any additional fertilizing, but if their leaves begin yellowing, apply a liquid fertilizer at half strength.

▪ If you don't use a prefertilized mix, or use a sterile planting medium like vermiculite, feed tomato seedlings weekly with a liquid fertilizer high in phosphorus and potash but low in nitrogen. You can also use manure tea (made by diluting dry cow manure in water), or compost tea. Remember that too much nitrogen is harmful to tomato seedlings, making them pale and leggy. Cut down on or stop fertilizing if you notice this reaction.

■ Never let seedling soil dry out—keep it moist but not sopping wet. A good way to test is to push a piece of cardboard half an inch into the soil. If it doesn't feel damp when you pull it out a few seconds later, it's time to water.

■ Don't water seedlings from the top so that water touches the plants. If water touches the plants directly, the fungus disease damping-off may occur. Either be very careful when watering from the top, or soak the seed flat briefly in a pan of water until the soil is subirrigated (watered by capillary action from below).

■ If damping-off of seedlings does occur despite all your precautions, there are commercial preparations widely available at garden outlets to combat it.

Tomatoes should ideally be transplanted at least once while indoors, many experts say. Some gardeners even transplant tomato seedlings from flats to a two-inch pot when they are 2½ inches high and then again to a larger pot when they reach about four inches. I've never found much difference between tomatoes raised with one or two transplantings indoors, but my plants raised without any transplanting are invariably weaker and more spindly than those that have been transplanted once. Tomato seedlings not only seem to like the less-cramped root conditions in a new, larger container, but welcome a new soil mix containing fresh nutrients. **TRANSPLANTING SEEDLINGS INDOORS**

Be sure that your hands are clean when you handle tomato seedlings during transplanting to avoid transmitting diseases, and don't smoke—tobacco mosaic disease can be spread in this way. Then follow these simple steps:

1) Choose a transplanting container from those mentioned in Chapter Five. A three-inch pot is good for standard variety tomato seedlings and a four-inch pot for hybrids. (Tests by Burpee show that hybrid tomato seedlings in four-inch pots produce earlier and yield more heavily than those raised in smaller pots.)

2) Fill the transplanting container with a new batch of the same planting medium used for the seed flat. (If you transplant into Jiffy 7 Pellets or BR8 Blocks, of course, you will need no soil.) Growers often fill containers with a slightly richer mix of compost, sand, and loam in equal parts on transplanting, but all prefertilized soil mixtures have ample nutrients.

3) After watering them carefully, remove the four-inch-high seedlings one at a time from the seed flat. A flat stick, a spoon, or even an apple corer can be used here, but it is best to take a dull knife and cut each plant out with a block of soil

so that its roots are intact. Then plant the seedling in its new container. Hold the transplants by the rootball or the leaves, not the stem—the slightest pressure can break tomato seedling stems.

4) Keep the transplanted seedlings in the same sunny window (or under lights) and at the same temperatures as you did the seed flat. Fertilize and water in the same way, taking care never to let the soil dry out.

5) Whenever the plants appear to be getting too thin and leggy, pinch them back by nipping out the growing tip with your thumb and forefinger. This will make them stockier and result in more blooms per plant.

Tomato seedlings at the proper stage of growth for transplanting into larger growing containers.

HARDENING-OFF SEEDLINGS

When your plants are eight to twelve inches high and seven to eight weeks old, they'll be ready for transplanting to the garden. A week to ten days before this, begin to harden them off by setting them outdoors and cutting down on watering. Just put the plants outside in full sun on favorable days and bring them in at night. Do this gradually; a half hour the first day, a little longer the second until they are outdoors almost all day. You can also transfer your plants to a cold frame to harden them off. (Instructions for building one are included in Chapter

Seven.) Hardening-off adjusts plants to outdoor conditions, toughening the tender tissues that they developed indoors and reducing injury from unexpected temperature drops. Giving the plants less water slows down growth and enables them to better withstand the shock of transplanting to the garden.

Instead of seeding tomatoes into containers prepared at home, you can use one of the many starter kits available. A million people a year use this Punch 'N Gro kit, which is preplanted with tomato seed, prefertilized, and only needs to be watered until the plants are set outside. Chief drawback is that not many varieties are offered.

Chapter Seven

PRECOCIOUS TOMATOES:
GAMBLING FOR THE EARLY ONES

Every true vegetable gardener is a gambler trying to harvest crops earlier than the experts advise, and early tomatoes are the most prized of precocious crops. There are several ways to raise early tomatoes from seed that will put fresh fruit on the table four to six weeks before most gardeners have it and at a time when it isn't too hot to harvest lettuce, radishes, and other salad-bowl vegetables. When using any of these methods observe the same rules given regarding seeds, soils, containers, and seedling care. However, try to choose seed of an early-bearing determinate or bushy variety (see Appendix I), not a main-crop tomato. Some early-planting stratagems follow:

EARLY
TRANSPLANTS:
SOW SEED EARLY,
SET PLANTS OUT
EARLY

Begin by sowing your tomato seeds inside *three to four weeks earlier* than you normally would. Plant the seeds in individual pots and tend them like any tomato seedlings, especially in regard to chilling and hardening-off the plants as described in Chapter Six.

Set the hardened-off plants out in the garden three to four weeks earlier than usual. Observe all outdoor transplanting tips noted in Chapter Eight, but try to set early tomato plants in an extra-rich, warmer soil. One or several of the following tactics will help:

▪ Dig planting holes about three weeks before setting out plants and fill them with decaying (heat-generating) organic matter such as manure, leaves, compost, or fish scraps.

▪ Remove leaves from an area heavily mulched the previous autumn and plant tomatoes in this space after enriching the soil with well-rotted manure.

Ultra Girl VFN is an early-starting type with excellent disease tolerance.

- Cover the planting area with black roofing paper or black plastic, which will absorb sunlight and heat the soil in the daytime and serve as a blanket at night.
- Fill plastic bags with water and lay them over the soil in the planting area. They'll warm up during the day and heat the soil on cool nights.
- Set your plants in a trench. Recent experiments at the Agricultural Research Station in Weslaco, Texas, show that tomato plants set out in cold weather do much better if they are planted in a shallow trench six inches deep and four to seven inches wide. Plants set in planting holes in such trenches enjoy higher nighttime moisture due to less evaporation. The Texas tomatoes raised in this manner yielded one third more fruits than conventionally planted tomatoes and matured three weeks earlier.

Small Fry, an All-America winner usually grown in pots, can be set out as an early plant, too.

■ Make a raised bed for early transplants the previous autumn. Soil in a raised bed never becomes waterlogged, warming up and being ready to plant weeks before ordinary garden soil can be worked in the spring. Raised beds are built in two ways. One method calls for laying railroad ties or foot-wide boards around the chosen area and filling the enclosure with rich topsoil. Or you can shovel topsoil over alternate layers of leaves, cornstalks, hay, and similar organic wastes until the board-enclosed area is filled.

When planting early tomatoes try to take advantage of the microclimatic conditions in your yard, too. Always keep a thermometer in the garden so that you know the exact temperatures there—they might be considerably different than in the general area. Plant where the ground slopes to the south, increasing the amount of sunshine plants will receive. Never plant where the ground slopes to the north. You can even create southern slopes where none exist.

Never set early transplants into soil that is mulched with organic materials— this will only keep the ground from warming up by reducing solar radiation into the soil. If your vegetable garden is mulched, push away the hay or leaves and

Springset, an early type, yields abundant large fruits.

Coldset, a tomato for early planting and direct seeding, compared with fruits of a small cherry-type plant.

plant the tomatoes in the bare earth. **Replace the mulch when warm weather** comes.

Always stake large-vined early transplants (see Chapter Nine) to protect them from wind damage, and don't fail to protect all early tomato transplants from cold and frost. Use one of the protection methods listed at the end of this chapter.

EARLY PLANTS
FROM CUTTINGS

If you have large tomato plants grown in the greenhouse over the winter, it is better to take cuttings from them and plant these out in the garden rather than the large plants themselves. Plants made from cuttings will be hardier and bear fruit early. To make a cutting, begin about three months before you would set plants outside. Choose a side branch six inches long. Cut this on a slant and root it in moist sand immediately after dipping it in a package of commercial rooting hormone. Keep the sand moist and when the cutting roots, transplant it to a peat pot filled with an artificial soil mix. From then on treat the plant as outlined above in Early Transplants. . . .

SEEDING EARLY
TOMATOES INTO A
COLD FRAME

A cold frame is a large rectangular bottomless box with a glass top that is placed over the ground outside. A small greenhouse in principle, it is airtight and heated only by the sun. You can get a jump on the season by seeding tomatoes directly into a cold frame about three to four weeks before the average last frost in your locality. The plants can either be grown in place until they yield fruit, or transplanted to the garden when they're eight to ten inches high. As noted, indoor-grown tomato plants can also be hardened-off in a cold frame, so it is a useful gardening aid to have.

Cold frames cost about $50 for a 3-by-6 size, but you can easily build one from inexpensive materials you might have at hand or can scavenge elsewhere. There are many plans available for elaborate types using redwood and other expensive supplies. (Your county agricultural agent will send you a free USDA pamphlet giving exact specifications for several.) Yet a simple frame to suit the needs of most amateur gardeners can be made with cheap lumber and ordinary storm windows or an old glass door. First, measure the storm window you have and buy or scrounge enough 2-by-8 second-grade wood to enclose it. Then measure off and enclose with string a rectangle in a space facing south on high ground where the cold frame can stand permanently. Be sure that the soil has good drainage; if it doesn't, dig out the bottom soil and replace it with coarse rock and then gravel before refilling.

For soil in any cold frame with good drainage, dig out about a foot of earth from the top of the string-enclosed area and refill the space with six inches of rich, light soil. Then place the four boards for the frame against the sides of the excavation. The tops of all the boards should now be exactly the same height above the ground (about two inches) and must be level. Finally, just butt the boards together at right angles, nail them together solidly with three-inch-long nails, and top them with the storm window. With six inches of soil removed from the enclosure, plus two inches on top, this cold frame will give your plants eight inches worth of headroom to grow in, but if a taller frame is desired, simply use larger boards. I've often used boards a foot high (2 by 12s). If you want a wider cold frame, use two storm windows, or more, and additional lumber—there's no limit on how big the structure can be. The trick in the whole operation is to make sure all your measurements are exact so that the structure is airtight when you place the storm window on top. Remember *not* to treat the wood with chemical preservatives—these can infuse plants. Even inexpensive woods will last up to ten years before needing replacement.

Many gardeners build cold frames as above and use heavy-grade, clear plastic sheeting as a cover in place of glass, anchoring the plastic with boards so that it isn't blown away and pulling it back when ventilation is needed. Others substitute concrete blocks or bricks for the wood used in building the structure. For additional reflected light and heat, the inside walls of frames are often coated with silver paint.

Equip your cold frame with a thermometer on the inside. When temperatures drop very low, bank soil, hay, or sawdust up against the outside of the frame and cover the glass with a quilt, blankets, mats, straw, black plastic, or similar insulating material. When the weather is hot (over 70° F.) during the day, prop the front end of the storm window up on bricks to provide ventilation for the plants and prevent them from baking to death in a hot oven. You can also shade the cold frame with boards during unusually hot weather, or use a mist spray to humidify and cool the air. Do all watering in the morning so the plants will dry off by night.

Use only seed treated against damping-off fungus when seeding tomatoes directly into a cold frame. As an added precaution, sterilize the soil by pouring five or six gallons of boiling water over it a day or so before planting. Sow the tomato seed about a quarter inch deep, following the same procedures as for seeds planted directly into the soil outside. (See Chapter Four.) Water carefully, taking care not to get the soil too wet during periods of low temperature or on cloudy days. If you are growing the plants in place and don't intend to transplant them to the garden, remove the glass entirely when average day and night temperatures reach 70° F. The plants can then grow as high as you like. Fertilize and cultivate as you would for any tomato plant growing outside. (See Chapter Ten.)

SEEDING EARLY
TOMATOES INTO
A HOTBED

A hotbed is little more than a heated cold frame. At one time fresh horse manure generated the heat, but today relatively inexpensive electric soil-heating cables designed for outdoor use are generally employed. For a 3-by-6 hotbed, you'll need approximately forty feet of a cable rated at about 200 watts.

Build the hotbed the same way as a cold frame, but add an eight-inch board below ground level, and be sure to place the frame near an electrical outlet. Lay the heating cable in loops on a two-inch bed of vermiculite and cover it with one inch of soil. Then place a piece of wire mesh over the dirt to protect the cable from accidental damage and cover this with six inches of soil mixed with humus and two cups of a complete fertilizer such as 5-10-5.

Tomato seeds can be sown in hotbeds even earlier than in cold frames—a full eight weeks before the average last frost in your area. Sow seed just as for a cold frame and care for the plants in the same manner. Be particularly careful about ventilation in a hotbed, though. About two weeks before setting your plants out in the garden from the hotbed (hotbeds aren't generally used to grow tomato plants to maturity), turn off the heat in the bed, using it as a cold frame from then on and gradually hardening-off the plants. Additional information about hotbeds can be found in the free USDA pamphlet ⚹445 *Electric Heating of Hotbeds,* available from the U. S. Department of Agriculture, Washington, D.C. 20250.

A DOZEN WAYS TO
PROTECT EARLY
PLANTS OUTSIDE

Unless they are raised in a cold frame or hotbed, all tomatoes transplanted to the garden before the usual planting date in your area must be given protection. Try one of the following protection ploys, or devise a method of your own based on these:

▪ For centuries gardeners have been using cloches (bell-shaped glass covers) to protect plants. Hotkaps are the modern version of these. Hotkaps are inexpensive miniature hothouses made of weather-resistant wax paper that protect plants against the elements. Place Hotkaps over all your early transplants. They should be kept on the plants at night and removed when the weather warrants it. Hotkaps come in several sizes: Regular, 6 by 11 inches; King Size, 9½ by 9 inches; and Super-Titanic, 14½ by 18 by 12½ inches. Larger sizes are best for tomatoes.

▪ Large paper grocery bags are sometimes substituted for Hotkaps and slipped down over tomato stakes to cover early plants. Seal the hole made in the top of each bag with adhesive tape to make the fit snugger and prevent the bag from tearing. Weight the bottom of the bag with stones or soil to keep out cold air. Use the sacks whenever excessive sun, frost, or winds threaten the young, tender plants. Pull them up on the stakes when the weather is good.

■ Gallon-size plastic jugs with their bottoms cut out make excellent miniature greenhouses to protect early plants. You can cut these in half, too, so that just the top half has to be lifted off when the weather is pleasant. Mound soil under the jugs as the plants grow and the plants will never touch the top of the jugs.

■ Large-size coffee cans and tomato cans—the bigger the better—afford excellent plant protection, especially if both tops and bottoms are cut out and the top hole is covered with a small pane of glass. Use them as you would plastic jugs above.

■ You can build a "greenhouse" for a row of early tomato plants out of old storm windows standing lengthways—just dig the storm windows about eight inches into the soil on all sides of the row, seal each space where the windows join with tape, and top the "greenhouse" with either more storm windows or clear plastic. A simple wooden frame built over a row of plants and covered with clear plastic will serve the same purpose. Or you can make a plastic tent "greenhouse" by draping clear plastic over tall wire wickets placed in the row, clipping the plastic to the top of each wicket with clothespins, and holding the plastic in place at ground level by covering it with soil.

Early-tomato plant growing outdoors protected from frost by a paper hotcap.

■ Tall, open-top cylindrical cages made of concrete reinforcing wire can be placed over each early transplant and covered with clear plastic for protection. Later, the plastic can be removed and the cages used as stakes for your plants. (See Chapter Nine.)

■ For small plants, slotted bushel baskets covered with clear plastic make good protectors. So do bales of hay placed around a plant or two and topped with clear plastic or a sheet of glass.

■ Wire-mesh plastic sheeting, sometimes used for poultry-house windows, is often fashioned into tepees around individual early transplants. To make one, drive three long poles into the ground around a single tomato plant and tie them together at the top; then simply cut the plastic sheeting so that it fits the framework and use thumbtacks to close the flaps together. In really cold weather drape a blanket over the tepee for extra warmth.

It is important to remember that adequate ventilation must be provided with any covering used for early plants. This is particularly important in areas where cold or cool spring weather is occasionally broken by temperatures equal to a hot summer day. Plants can be broiled alive under tight coverings on such days. Either be sure to provide small openings that will let excess heat escape on warm spring days (and can be sealed when it gets really cold), or else go out in the garden and remove all coverings, replacing them at night. It's also a good idea to introduce a few beneficial insects such as ladybugs under any shelters you make. They will devour those harmful bugs that tend to congregate in warm places in early spring.

MORE TRICKS TO PROTECT EARLY TOMATOES AND HASTEN RIPENING

Hormone Sprays. These will insure fruits two to three weeks earlier if used on plants set out in cold weather. At flowering time, when night temperatures fall below 59° F., tomatoes will not set fruit in their flowers. Flowers will also fall off because of short cloudy days without much sunlight. What hormone sprays do is to travel through the plant and prevent the formation of the cutting-off or abscission layer that forms between bud and stem; it is the abscission layer that causes the flower to drop before the fruit is set. The hormone spray makes the fruit set and holds it on the plant. Treated plants will also produce more (sometimes twice as many), larger, and meatier fruits. In fact, treated fruits will be seedless because male pollen wasn't used, which is a blessing to those who like beefsteak tomatoes and minor tragedy to those who like types with abundant seeds. Hormone sprays are inexpensive and widely available. Blossom-set, Sure-Set, and Duraset are three good ones. Carefully follow all directions and spray into the face and not the backs of the flowers. Incorrect use, or application where an excess of nitrogen has been used to fertilize plants, may cause premature softening of tomatoes and puffiness or large hollow spaces in the fruit.

Root Pulling. To hasten ripening of tomatoes after a few on the vine have reached full size, grasp the main stem at ground level and pull upward until the roots snap. The largest fruit on the plant will ripen faster. The plant, which will wilt a bit, will quickly recover with no permanent damage. Try this on only one or two plants until you get the hang of it.

Picking Fruits Early. If you pick the first red tomato on a vine before it is fully ripe, this will hasten ripening of other fruits on the plant.

Water-spraying Plants. Water spraying serves two purposes: 1) A light morning sprinkling of water (of at least air temperature) for a couple of hours during a frost sometimes prevents plant tissues from freezing. According to Cornell University botanists, you should apply 1.10 inches of water an hour just as soon as the temperature falls near 32° F., wetting the plants at least once a minute until no more ice remains on them. 2) Water spraying helps plants set fruit during unexpected early hot spells. Spray their foliage and try not to wet the blossoms. Spraying can even be done at night if the plants dry off in an hour or so.

There's really no limit to the ways you can protect early tomatoes. In closing, consider the ingenuity of a determined tomato gardener whose early plants were hit by a severe, unexpected frost. He quickly constructed plastic tepees over the plants and saved most of them by placing an ordinary vaporizer inside the tents and leaving it on for a few hours until they revived.

Chapter Eight

THE BEST AND THE BRIGHTEST:
SETTING OUT MAIN-SEASON TOMATOES

Unless early tomatoes are your goal, tomato plants shouldn't be set outside until after the average last frost date in your locality, *and when both air and soil temperatures average at least 60° F. over twenty-four hours*. Night temperature is just as important as day temperature. An unprotected plant put out before temperatures are above 55° F. for several hours every night will simply sulk and refuse to set fruit. Cold nights are also the most frequent cause of blossom drop in spring. In this case, the blossoms emerge, but the pollen that germinates them takes so long growing down the pistil to the ovary in the cold that the blossoms drop off before they are fertilized. If you don't plant at the right time, the only way to protect against this is to use the protection measures recommended for early tomatoes in the last chapter.

The length of the growing season (the number of days from the average last spring frost to the average first fall frost) in your area must also be considered when planting tomatoes. Fruiting time in all gardening catalogs is calculated from the day plants are set out, not from the day seeds are planted inside. Thus, in an area with a growing season of ninety days beginning on June 1 and ending on August 31, it would be foolish to set out a Burpee Big Boy plant (eighty days to fruit) on June 15. Earlier planting or an earlier variety would definitely be called for.

It is a simple matter to determine your tomato planting date by consulting the local weather bureau, state agricultural extension service, a plant dealer in the vicinity, or an experienced local gardener. Planting when there is no danger of frost and after the soil warms up has one distinct advantage. Although the plants won't bear as early, they will almost always bear more and longer than plants set out under unfavorable conditions. Various studies show that tomatoes transplanted early aren't as healthy as their main-crop relatives and that cold weather checks

Tomato plant at proper stage of growth for transplanting into the garden.

the production of fruit buds so that they don't bear as much no matter how well protected.

A final tip: Folklore has it that the best time to plant tomatoes is when the moon is waxing, or "coming up." One explanation holds that just as the tides are controlled by the moon's pull, so is the upward and downward movement of moisture in the soil. Thus a waxing moon, one growing full, would provide more moisture for seedlings. A second theory postulates that it is the *increased light* of a waxing moon that helps plants. Still another conjecture has it that *electrical voltage* increases when the moon is waxing, thus aiding plants. You can take your choice, or refuse to believe any of this, but planting tomatoes when the moon is waxing certainly can't hurt them. . . .

PLACES TO PLANT TOMATOES

Be sure to review the basic tomato sunlight and soil needs given in Chapter Three if you haven't already noted them. See Chapter Fourteen, too, for many specific space-saving ways to plant tomatoes, including the use of patios, window boxes, planters, flower beds, and even a novel plan to utilize that great green desert, the front lawn. But wherever you put your tomato patch, try to choose an easily accessible place close to the house. Make certain there is easy access to water so that a hose or buckets of water won't have to be dragged a quarter of a mile across the back forty. Garden tools and supplies should be stored nearby, too.

Locate the tomato patch away from any heavily traveled road bordering on your property—exhaust fumes stunt plant growth and often kill tomato plants. A fence will help if there is no choice but to plant by a highway, but it won't be totally effective. Fences will also afford some protection against animal trespassers, wild or domestic, although many animals manage to bypass the best-built barriers in one way or another.

Finally, try never to plant tomatoes in the same area of the garden every year. Rotate crops, so that diseases and insects carried over in the soil from last year won't be as much of a problem. In fact, never plant tomatoes in a spot occupied by any member of the *Solanum* family (peppers, potatoes, eggplants, okra, etc.) the previous year. It is true that this again is an ideal rule (I have grown tomatoes in the same spot for five years at a time without any trouble) but if you're able to follow it—and most gardeners are—why take chances?

AVOIDING TREE ROOTS

Keep tomato plants forty to fifty feet away from shallow-rooted trees such as maples, elms, willows, and poplars. (Deep-rooted trees like oaks present no problem.) The feeder roots on trees reach out in the soil as far as their branches do in the air and can rob garden soil of moisture and nutrients. If tomato plants can't be kept at such a distance, due to space limitations, dig a narrow trench about three feet deep between the garden and the trees bordering it. Line one side of the trench with galvanized sheet metal or heavy plastic and refill the trench with soil. It will take years for the roots to work their way through this barrier and the root-pruning won't hurt the trees. Do the same with any large shrubs bordering your garden.

It is especially important not to plant tomatoes near black walnut trees, as the roots of this species produces a substance called juglone that is very toxic to many tomato varieties. Juglone is exhuded as far out as black walnut roots extend. You can try digging a ditch between black walnuts and the garden, but ditching doesn't always work. Juglone, which causes wilting and dwarfing of tomato plants, is not present in dead black walnut trees, or woodchips, bark, sawdust, and leaves from the trees, so these can be used in the compost pile as a mulch.

TOMATO POSITION
IN THE GARDEN

Whether you plant in a novel location or in a no-nonsense "kitchen garden," several rules should be followed when planning the position of the tomato patch. First, since tomato plants grow quite tall compared to most vegetables, always place them where they won't shade smaller growing crops. If garden rows run east and west, for example, plant the tomatoes on the north side of the garden. The only crops that should be planted behind tomatoes are corn, pole beans, and possibly cucumbers, peas, and other crops that grow on tall trellises. The smallest determinate or non-staking tomato varieties, however, can be grown in front of many crops.

Try to plant tomatoes on a southerly or southeasterly slope where the springtime sun will fall a little more directly than on a level surface. This will raise the temperature and get the tomatoes off to a better start. Locating plants on a sloping site will also provide the air circulation so important to tomatoes, because as air cools, it flows downhill, creating a current.

If you can find a sloping southerly spot for your plants against a garden wall, house, or garage, so much the better. One way to simulate a garden wall is to plant a dense row of pole beans behind the tomato rows.

As noted, tomatoes grown in sunny exposures generally taste better and are higher in vitamin C content as well. Protected from wind and warmed by a wall they will bear earlier and longer, as much as eight weeks longer, than other tomato plants. By planting them so, you actually create a miniclimate or microclimate in your yard equal to a climate about 250 miles farther south.

SOIL WORKABILITY   Before planting tomatoes make sure the ground is not too wet to work. Take soil samples at the surface and a few inches below it. If either sample sticks together in a ball and doesn't readily crumble under slight pressure, then the soil is still not ready to work. Another way to test for moisture is to insert a spade into the soil. If soil sticks to the spade, it is too wet for planting.

LIMING TOMATO SOIL   Remember not to lime tomato soil unless a soil test shows that liming is necessary. Soils in a high state of fertility do not need liming. The only other reason for liming besides raising the soil's pH level (see Chapter Three) is to help prevent blossom-end rot. Blossom-end rot—a condition where the free end of the tomato turns black—is partly caused by a calcium deficiency in the soil. This can be corrected by the application of ground dolomitic limestone just before planting. If you must lime for this reason, and haven't had a soil test made, just lightly dust the soil with dolomitic lime a few days before planting, using about five pounds per hundred square feet. Then rake the lime thoroughly into the soil.

PREFERTILIZING TOMATO SOIL   Fertilize the garden a few days before setting out tomatoes. If the soil has been tested, follow directions supplied by your state agricultural agent or other testing service. If not, use compost, manure, or any other organic fertilizer in liberal amounts—they can only benefit the plants—or use commercial fertilizer at the rate specified on the bag. A general purpose 5-10-5 fertilizer (5 per cent nitrogen, 10 per cent phosphoric acid, and 5 per cent potash) gives good results, but there are many specifically formulated tomato preparations on the market. Stay away from fertilizers that have a high nitrogen content—they will only stimulate green growth at the expense of fruit. To make your own tomato fertilizer, thoroughly mix eight tablespoons of cottonseed meal, eight tablespoons of superphosphate, and four tablespoons of sulfate of potash into each bushel of garden soil. This will prove a little cheaper than buying specific tomato preparations.

In applying fertilizer most gardeners spade the plot first, broadcast the fertilizer by hand or with a spreader, and then rake the soil two or three times to mix in the fertilizer. But it is better to apply dry fertilizer in circles or bands around each tomato plant, each circle a few inches deep and about four inches from the plant base. Tomatoes have a need for phosphorus as the roots develop. When fertilizer is simply broadcast and worked into the soil, a lot of the phosphorous is locked up by the soil and isn't immediately available to the plants. Concentrate the fertilizer in a ring and the plant immediately gets what phosphorous it needs, even though much of it is locked up.

When sowing tomato seeds directly into the garden, be careful not to place commercial fertilizer directly on the seeds where it will burn seedling roots. To avoid this, tie a string tautly to two stakes where the row of seeds is to be planted. Then dig furrows three inches deep and three inches away from the string on both sides. Spread the fertilizer in the furrows and cover with soil. Sow the seed directly under the string.

Here are some step-by-step tips for the actual transplanting of tomato plants into the garden that will both hasten and increase their yield:

SETTING TOMATO TRANSPLANTS INTO THE GARDEN

Removing Blossoms. Prior to setting out plants, remove all their fruits and blossoms. This may seem harsh, but if they are allowed to grow, the plant root systems will never develop properly. Retaining fruits and flowers can only be justified if you value a few early fruits enough to sacrifice the whole plant for them.

Weather. An old gardening rhyme instructs: *This rule in the garden do not forget: / Always sow dry and set wet.* Tomato *plants* (as opposed to tomato *seeds*) should whenever possible be set out on a rainy or cloudy day in order to lessen transplant shock for the plants. Evening is a good time to set plants and just before a rain is best of all. Avoid planting on bright sunny or windy days—or at least shelter the plants for several days with Hotkaps, strawberry boxes, paper cones, or similar protection if you must do so. (See Chapter Seven section on Plant Protection.) If the soil is bone dry, water it until it is slightly damp before transplanting.

Spacing. Tomato plants that are to be staked should be spaced about eighteen inches apart in rows three feet apart. Plants that will be allowed to sprawl on hay or another mulch should be planted three feet apart in rows four to five feet apart. Like most instructions, however, these represent ideal solutions. Standard tomatoes can be planted much closer together if need be and staked plants are often set 2 by 2 feet or even 1½ by 1½ feet apart without sacrificing much yield. Vigorous hybrids should be given two feet each way wherever possible. Many determinate-type tomatoes, which grow only two to three feet high in an upright manner, need no more than a foot between plants. There is plenty of room for experiment here.

Removing Plants from Pots. Transplants in peat pots, cubes, and blocks can of course be set into the planting holes just as they are—with the tops of their containers set at least an inch below the soil. Plants in other containers should be removed with as little root disturbance as possible. Thoroughly

TRANSPLANTING TOMATO PLANTS
TO THE GARDEN.

Water the seed flat or any container thoroughly.

Dig a hole that will accommodate all of the plant's root ball plus a little of the stem.

Fill the planting hole with water and let it drain off.

Remove the plant from the flat with as much root as possible and place in the hole.

Fill dirt in the hole, taking care to firm it around the roots. Water the transplant well.

wet the soil in plastic, paper, or clay pots and the root balls will tip out easier. Just tap them out by holding the pot upside down and knocking the bottom of the pot with the palm of one hand, while holding the top of the pot and the stem of the plant securely in your other hand. If your plants are growing together in a large flat, use a sharp knife to cut out squares of soil containing each plant. Be sure to cut down to the bottom of the flat, so that you retain as many roots as possible.

Planting Depth. Dig generous holes a few inches deeper than the rootball of each plant. In this way you will be burying part of the plant stem; feeder roots form anywhere along a tomato stem, and plants buried deeply will form healthier, denser root systems. They will also be better anchored against possible wind damage. When plants are very scrawny and leggy, bury even more of the stem in the ground. Dig a hole to accommodate all but about three inches of the leggy plant. Or make a small trench long enough for the stem and lay the leggy tomato plant in it on its side. Very gently bend up three inches of the tip and then cover the rest of the plant as if it had been planted in an upright position. Don't worry about sacrificing indoor growth. The leggy plants will grow better and soon catch up with plants towering above them.

Filling In. Once the transplants are in their planting holes, press the removed soil firmly against them, leaving a slight one-inch saucerlike depression around each plant to hold water. It is important the soil be packed tight against the roots.

Watering and Starter Solutions. Always water tomato transplants. Many gardeners feed their young plants at this time with a "booster" or water-soluble fertilizer. Or they simply dip the plant roots in the booster solution before planting. Transplants that receive starter solutions are generally more vigorous and mature earlier, their root systems becoming established more quickly. There are numerous starter solutions on the market for tomatoes. Usually they are high analysis fertilizers such as 6-18-6, all of them high in phosphorous as opposed to nitrogen and potash. But you can make a good starter solution by adding two tablespoons of 5-10-5 fertilizer to a gallon of water. Water each plant with about a pint of this solution after planting.

See that your young plants get plenty of water until they are established, but don't mulch plants until the ground warms up (see Chapter Eleven) and don't let the ground become soggy around them. After the transplants are established in a few weeks, it is a good idea to stretch the interval between watering, putting the plants under slight stress until fruit sets. From then on water at the rate of about one inch per week.

A new device, Burpee's Automator, takes the worry out of watering. The Automator waters, feeds, and warms soil for early growth. This heavy plastic tray is set at the base of a tomato at planting time and you simply water into it. A box of three costs $3.95 and the Automators can be used from year to year.

Cutworm Collars. After setting plants out, make a stiff cardboard or paper collar about six inches wide for each plant. This is easily done by forming a circle with cardboard or paper and holding it in place with a paper clip. Sink the collar into the soil around a plant to about half its depth. Collars can also be made from milk containers, open-ended paper cups and tin cans opened at both ends. Any similar device will prevent cutworm larvae from crawling up the stems of young plants at night and feeding on them.

Plant Protection. Remember to make provision for protecting plants if there is any chance of frost in the area after planting time. (See Chapter Seven, Plant Protection.)

Markers. When planting different varieties of tomatoes, be sure to put identifying markers near each plant so that you know which is which. Ice cream bar sticks are good for this purpose. Variety names can also be written on the plant stakes.

Chapter Nine

TO STAKE OR NOT TO STAKE:
STAKING, PRUNING, AND PLANTING
COMPANION CROPS BETWEEN TOMATOES

W hether to stake or not to stake tomatoes isn't any easy question to answer, for there are virtues in both letting plants sprawl naturally, as market growers usually do, and tying them to various supports, which is the practice of most home gardeners.

Contrary to popular belief, tomato plants allowed to sprawl will almost always yield more fruits than plants *staked in the traditional ways.* (Twenty fruits as opposed to about twelve are the average.) Obviously, there will also be little or no work involved in tying and pruning when plants are allowed to sprawl. Unstaked plants generally produce less sunscalded fruit, too; their fruit tends to crack less, and there is less likelihood of blossom-end rot on their fruits in most seasons. Due to their heavier unpruned foliage they are less susceptible to the ravages of drought. Finally, fewer plants are required for the same total production than with staked plants, and unstaked plants are easier to protect during the cold weather toward the end of the growing season.

On the other hand, the new ring-culture tomato-staking methods yield *more* fruit than unstaked tomatoes. And tomato plants staked in any manner always fruit and ripen a little earlier (though not much earlier) than unstaked ones. Staking also results in fruit that is cleaner, free of ground spots, and less subject to rotting and damage from slugs. Fruit on staked vines is a little larger on the average, and there are definitely fewer small tomatoes on staked vines. Perhaps most importantly, fruits on staked tomatoes are easier to spot on the vine, can be easily picked without much bending, and make for a neater garden. If you want to grow a large number of varieties where space is a problem, almost twice as many staked plants can be grown in a given area.

Bushy determinate tomato varieties definitely don't need to be staked and, in any event, should *never* be pruned to train them to a stake. These stout-stemmed plants, usually early types, grow no more than three feet tall in an upright form and stop growing when fruit begins to set. Their upright posture makes staking unnecessary and if they are pruned their yield will be considerably reduced. Semideterminate varieties, which grow a little larger, often don't have to be staked, either, and neither should they be pruned. It is the indeterminate varieties—tomato plants with stems that grow indefinitely in length and suffer no ills from pruning—that you'll have to decide whether to stake or not. Appendixes I and II indicate many varieties that are determinate, semideterminate, and indeterminate. Some of the better non-staking types include: Bonanza; Campbell's 1327; Chico; Fireball; Galaxy; Heinz 1350 (semideterminate); Jetfire; New Yorker; Pearson ⅌9; Porter; Primabel; Rocket; Roma; Small Fry; Spring Giant Hybrid; Springset; Stakeless; Swift; Tiny Tim; Walter; and Yellow Tiny Tim.

IF YOU DON'T STAKE TOMATOES

Quite a few gardeners with lots of space available elect to let even their vigorous indeterminate tomato plants sprawl naturally. If you do so, try to give each plant at least fifteen square feet of room. (Space the plants 4 by 4 or 3 by 5 feet apart.) But don't neglect the plants just because they are sprawling naturally. Examine them periodically for slug and hornworm damage and apply corrective measures where needed. (See Chapter Twelve.) Above all, give the fruits some-

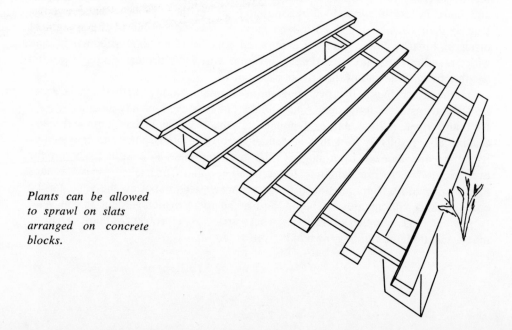

Plants can be allowed to sprawl on slats arranged on concrete blocks.

Unstaked, unpruned plants usually bear more than staked, pruned specimens, but fruit is subject to soil rot unless mulched.

VF 10, an upright, determinate plant that needs no staking.

thing to rest on so that they don't develop ground rot and blemishes. Either organic or non-organic materials can be used. Several suggestions follow:

- Construct a low platform made from cinder blocks and long sticks or slats. Place the cinder blocks at intervals along the sides of the tomato patch and lay the slats across from side to side. Let the plants grow naturally over the slats.
- Lay a deep mulch of clean salt hay, baled hay, straw, grass clippings, buckwheat hulls, seaweed, or any similar organic material in the tomato rows for unstaked plants to sprawl on. Even tree branches piled next to the plants or the packing from an old hassock will keep them off the ground. The mulch should be at least six inches deep in all places. (See Chapter Eleven for a more complete list of good mulches and additional benefits of mulching.)
- Use inorganic materials like old newspapers laid in a thickness of six layers for protective mulches. The most popular of the inorganics is black plastic or polyethylene. This should be laid over the soil before tomatoes are planted. Buy 1½ mls. (.0015) thick black plastic and cover the prepared soil with it on a calm day. Around the edges of the plastic dig a shallow trench and bury about three inches of plastic on all sides. Then make holes in the plastic and set the transplants in them. The tomato patch will be weed-free and fruits will be clean.

TOMATO STAKING
BASICS

If you decide to stake your tomato plants, it doesn't make much difference whether staking is done just before setting out the plants or just after. I've never had a plant whose roots were damaged enough by staking at any time to do it any harm whatsoever. Care must be taken when tying tomato plants to stakes, though. Soft twine, plant ties, cloth strips, old nylon stockings, or raffia can all be used for ties. Tie the twine or whatever *tightly around the stake* two to three inches above a leaf stem; then loop it *loosely around the main stem* not far below the base of the leaf stem and fasten with a square knot or figure-eight loop. Try to keep any flower clusters away from the stakes, otherwise the tomatoes will be crowded between the stem and the stake as they enlarge and become injured or misshapen as a result. Check plants every four or five days to see if they need extra support. You'll probably have to tie plants every ten days or so, depending on their growth rate.

Staking Tip: Never bend tomato vines at sharp angles when tying them—they snap very easily. However, if you do break a stem and the pieces aren't completely severed from each other, join them together and tie them up with tape or string. You'll have saved the branch if its leaves don't wilt in a day or so and it will be back to normal in a few weeks.

When it comes to choosing the right kind of tomato vine supports, you're really only limited by your ingenuity. A dozen or so possibilities, some of which increase yield in tomatoes, are listed below:

Wooden Stakes. These are the most common of tomato support systems. The stakes are usually 1-by-1-inch boards six to eight feet long and made of cheap lumber which is coated with a wood preservative and often painted green. Wooden stakes are pushed, sledgehammered, or dug about one foot into the soil at a distance of three to five inches from the tomato plants. The portion underground tends to rot out every year or so, but when that happens the stakes can simply be hammered down deeper until they become too short to use.

Scrap lumber, furring strips—even straight tree branches—are all used for wooden stakes. Stakes do not necessarily have to be 1 by 1 inches, so long as they're strong enough to provide support. As for length, if you are growing very tall climbing varieties, you'll want stakes twelve feet high or longer. The worst wooden stakes you can buy are the spaghetti-thin bamboo poles sold in garden centers; these won't support healthy tomato plants. When staking with bamboo poles, use at least the one-inch diameter size. Nine-foot lengths in this size are sold by the Bamboo and Rattan Works, 901 Jefferson Street, Hoboken, New Jersey 07030.

Plants spaced two and a half feet apart and tied to individual wooden slats with plant ties.

Multiple Stakes. For heavy-vined tomato varieties, some gardeners sink two wooden stakes with cross supports next to each plant. The foot-long cross supports should be nailed to the main stakes at two-foot intervals *before* sinking them into the ground. In a similar fashion, you can make a boxlike support system for each plant from four stakes and cross supports. Multiple staking systems like these enable you to tie vigorous varieties in more places.

Pole Fences. Cross two bamboo poles, leaving two feet between the bases of the poles. Then tie the poles together about one foot from the top. Sink this cross into the ground and make another, setting it two feet down the tomato row from the first cross. Make enough crosses to fill up the row and then lay a long tree branch (or branches) down the top, so that it rests in the crotches formed by the top of the crosses. Set your plants at the foot of each pole and tie them as they grow up the poles. Narrow boards or straight tree branches can be used for the crosses instead of the bamboo poles.

Wire Fences. Here two stout eight-foot-high stakes are driven two feet into the ground ten feet apart. Lines of heavy wire are attached to them, each line two feet above the other. Five tomato plants can be planted at 1½-foot intervals along the bottom of the wire fence and tied or trained to the taut lines of wire as they grow. Space will be saved if plants are set on both sides of the wire fence, but the plants won't yield as well.

Double Wire Fences. You can plant a row of tomatoes between two wire fences set on each side of a row of plants and train the plants to both fences. Construct the fences the same way as for a single wire fence. Plants in a double-fence enclosure should definitely be mulched before the fences are built, as weeding is difficult under this system.

Wire and Slat Fences. Attach fifteen feet of sturdy concrete reinforcing wire to stout stakes spaced five feet apart. Sink this wire fence in the ground on one side of a row of plants. Erect an identical wire fence on the other side of the row. Then slide slats or sticks through the wireholed fencing on each side of the plants to give support when and where it is needed without any tying. The plants will simply lie on the slats that stretch from one fence to the other. Here again plants must be mulched due to lack of elbow room for weeding.

Other Fences. Any standing metal or wooden fence that is at least six feet tall makes a good support for tomato vines. A wooden picket fence is especially good; as the plants grow just tie them to the boards. Or you can fasten poultry netting to a picket fence, or run twine through the pickets for additional support.

Wooden Tepees or Tripods. Drive three seven-foot poles or branches about a foot into the ground two feet from each other, force them together at the top and tie the tops together with stout cord or wire. Set one plant in the middle of the tepee and prune it to a single stem, placing a six-foot wooden stake in the ground next to it. When the plant's main stem grows to the top of the stake, tie it to the tepee top. Tie all side branches to the three tepee poles. This system is especially good for vigorous tomato varieties.

Trellises. Store-bought wooden trellises can be set in the garden row, or the trellises might be made from lumber or tree branches uniform in size. Use heavier wood for the vertical supports than you do for the horizontal sup-

Tomato tepees, one plant at the base of each stick, make for an attractive, efficient staking system.

A single climbing tomato plant, like this twenty-foot tall Trip-L-Crop, will save garden space and yield hundreds of large fruits.

ports when building your own trellis and place the horizontal supports about one foot apart. The trellis should be at least six feet high and as long as is needed.

Pea Trellising. Support tomatoes like peas on a trellis. Set heavy six-foot-tall posts in the row every ten feet or so and stretch a line of heavy wire across the top from post to post. Tie heavy cords to the wire that will drape down to the plants. Attach each cord to the stem of a plant just above the ground. As the plant grows, keep it pruned to a single stem and twisted around the cord.

Odd Ones. If only to illustrate how gardeners have used their ingenuity in staking tomatoes, here are several unusual methods you might want to try:

▪ Cut off two sides of the wooden frame that supports the inside of large packing crates such as those used for refrigerators and stoves. Set what's left of the frame in the garden like an arch and grow plants along the sides of it.

▪ Make a tomato fence from old four-paned window frames with the glass knocked out. Just nail the old frames to posts sunk in the ground.

▪ Use sunflowers as living tomato stakes. Plant sunflowers close by at the same time tomatoes are set out. Let the tomatoes grow up three-foot wooden stakes at first. By the time the tomato plants reach the top of these wooden stakes, the sunflowers will be ready to use as natural stakes for the rest of the season. Tie the tomato stems to the sunflower stalk with soft cloth ties. Fertilize and water heavily to provide for the needs of both plants. Besides providing natural stakes, the sunflowers will shade the tomato plants and help prevent sunscalded fruit.

▪ In staking giant climbing plants—some of which grow over twenty feet high—use your imagination. One gardener trains his giant plants to wooden and aluminum extension ladders propped against his garage. No matter what kind of staking is devised for climbing plants—large stakes made by nailing boards together, long tree limbs, etc.—be sure that they're anchored at least three feet in the ground so that they won't topple over. Don't forget that you'll have to use at least a ten-foot stepladder to care for giant plants and pick fruit as they grow up into the stratosphere.

FREE TOMATO STAKES YOU CAN GROW IN YOUR BACKYARD

All the free stakes needed for tomatoes can be grown in your own yard if you plant shrubby willows like *Salix viminalis, Salix purpurea, Salix caprea, Salix discolor,* and *Salix gracilistyla.* If these bushes are cut back hard, eight-foot shoots can often be grown in one year, though it is best to leave the shoots on the plant for another year until they thicken and harden. Willow sticks root very easily;

they should be cut early in the winter so that they'll dry out before being used. Nurseries offering shrubby willows include: Zilke Brothers, Baroda, Michigan 49101; Waynesboro Nurseries, Waynesboro, Virginia 22080; and Girard Nurseries, Geneva, Ohio 44041.

Gardeners often underestimate how tall tomato plants will grow. They are left with vines that grow much higher than their stakes and frequently break at the top when loaded with fruit. It's easy to solve the problem. Simply add bamboo or wood extensions to the original stakes. Let the bamboo or wood extension overlap the original stake by at least a foot and fasten it with three pieces of wire to the stake. Make sure the wire is tight.

TOMATO-STAKE
EXTENSIONS

Wooden stakes with extensions are used to stake these vigorous Early Girl plants. Wire, stretched stake to stake, gives added support.

Tomatoes raised in wire cages or rings bear prolifically. Single plants in cages have yielded up to sixty-two pounds of fruit, and at the University of Maryland twenty-five plants grown in such corsets actually produced three quarters of a ton of tomatoes. There is some evidence that plants grown in wire cages receive the benefits of electroculture, each wire cage setting up a growth-stimulating electromagnetic field. Several popular ring methods follow:

Wire Cages or "Tomato Trees." This technique goes by many names—the "Chinese tomato ring," "tomato trellising," "corset growing," and "wire cylinder growing" are just a few. There are a number of approaches, but the easiest way to make a wire cage for a plant is to buy a piece of sturdy concrete reinforcing wire five feet high and fifteen feet long. Using an ordinary pair of pliers, form the wire into a cage or cylinder five feet high and five feet in circumference. Push the cage into the ground over a single tomato plant,

Wire fencing or concrete reinforcing wire about six feet high may be used to support a tomato plant by encircling the plant with the fence.

leaving the plant inside the cage. To insure that the cage is set far enough into the ground, cut out some of the cross wires at the bottom so that you can push the ends in at least six inches deep.

Other kinds of wire can be used to make tomato cages. Mesh hog wire or chicken wire are alternatives if they are supported with long stakes within the cage, but concrete reinforcing wire is best because it is rigid and self-supporting and because it has a wide six-inch mesh which enables you to reach in and pick the tomatoes easily. Smaller, assembled tomato cages (twenty-eight

Tomato cages like this one are available from many nurseries.

inches tall) can be ordered from several suppliers, including World Art and Gift, Department 4-3T, 606 East State Street, Westport, Connecticut 06880 at three for $6.75 or six for $10.95, including packaging and handling.

Tomato plants in wire cages should be mulched, for it is hard to weed them. Tie the plants to the wire with soft cloths as they grow, or just weave them in and out of the mesh, keeping all the branches within the cage. *Do not prune the plants.* When the plants reach the top of their cages, let them grow over the top and down the outside of the cages; that will make a total of ten feet of growing room for each plant by the time it reaches the bottom of a five-foot-tall cage on the outside. Be sure to feed tomatoes grown in this way often—at least once a month—using liquid fertilizer.

Wire cages also allow plants to develop naturally, providing shade for ripening fruit and reducing sunscald and cracking. They can easily be fitted with plastic to provide protection for plants in cold weather and last for twenty years or more. Recent experiments at the Overton, Texas, Agricultural Experimental Station have shown that a twelve-inch-high strip of roofing felt wrapped around the base of wire cages when early plants are set out will increase yield even more. The felt protects against wind and hail damage.

Japanese Rings. A relatively new cage-staking method, pioneered by Japanese and English gardeners, the Japanese ring-staking system is said to produce up to one hundred pounds of fruit on a single plant. The system is really an improvement on the preceding wire-cage method. Here the plants are grown *outside* the wire cage. First, fill the planting area that will be inside and just outside the cage to a height of two feet with six-inch layers of mulch, rich loamy top soil, more mulch, and more topsoil. Hold this pile in place with a two-foot strip of fine-mesh screen, and shape the top of the pile into a shallow saucer that can hold water. Then place a wire cage made from concrete reinforcing wire over it. Set four plants around the *outside* of the cage and tie them to the wire with soft cloth as they grow. Keep the soil in the center of the ring moist and add fertilizer to it every two to three weeks. Eventually, the plants will grow over the wire and down into the ring. They should be loaded with fruit.

Soil-aggregate Rings. If you want to try something really novel, these are the very latest thing in ring-staking methods. Here the wire cylinders previously described are placed over tomato plants set in a layer of rich soil with an inert aggregate such as ashes or vermiculite beneath it. Enormous secondary roots, which draw up great amounts of water, develop in the ashes, while dense fibrous roots that absorb nutrients are developed in the soil above. In order to encourage this kind of root growth, *the soil* is given frequent feedings with liquid fertilizer (but no water), while *the ashes* receive only water, which is introduced through a tube or hose so that it doesn't pass through the soil. Soil-aggregate rings produce tremendous crops of fruit. More about these innovative methods can be found in *Ring Culture* by Frank Allerton (Faber & Faber, 24 Russell Square, London W.C.1, England, $1.75).

PRUNING STAKED
TOMATOES

It might be considered heresy to say so, but in most cases, it doesn't really make much difference whether you prune tomatoes or not. As mentioned, the only exceptions to this rule are stakeless, determinate tomato varieties, or plants that are grown in wire cages; you'll only cut down on their yield by pruning. Tomato plants that are allowed to sprawl on mulch shouldn't be pruned much, either, unless they produce rampant green growth and no blossoms due to an overly rich soil, in which case they'll benefit from being cut back by a scythe.

It is staked indeterminate-type tomatoes—the really vigorous growers—that are usually pruned. But you can forget about pruning these, too. You'll grow all the tomatoes you'll be able to use just by letting your plants grow and tying them to their stakes. In fact, unpruned staked plants of any variety will yield *more* fruit over a longer period than pruned ones, for the suckers that are removed from pruned plants bear fruit if left alone. The main advantages of pruning vigorous plants are that they will be easier to stake and will yield *earlier* fruit. Even this last benefit is offset by foliage cover being diminished, with resulting sunscald damage to fruit. It should also be pointed out that the quality of fruit on pruned vines can be inferior to that on unpruned vines, for when leaves are removed, plants do not produce as much sugar and starch. On the other hand, it is a good idea to prune plants in a humid climate or during periods when there are heavy rains. In these situations dense foliage prevents vines and soil from drying and leads to fruit rot.

Should you decide to prune your plants, there are several methods to choose from. In following any of these techniques it is best to do all pruning with your fingers; if you use a knife or shears, you can transmit virus diseases from one plant to another. Pruning by hand is called "pinching out." Just grasp the small shoot you want to remove with your thumb and forefinger and bend it sharply to one side until it snaps. Then pull the shoot off in the opposite direction, which will prevent injury to the leaf axil or main stem:

Single-stem Pruning. Single-stem pruning yields fruit a week or two weeks before other methods. The English often raise their tomatoes this way. Here all suckers, or sideshoots, are removed from the plant and only one stem is al-

Staked, unpruned tomatoes yield prolifically.

lowed to develop and grow up the stake. Suckers are shoots that appear in the axils of the leaves (where the leaf is attached to the stem) at a 45-degree angle. It is best to check plants for suckers once a week and remove them while they are just forming; but this can really be done at anytime without harming the plants, even when the suckers are over a foot long. Be certain not to disturb the fruit buds, which appear just above or below the points where leaves are attached to the leaf stem.

In single-stem pruning also remove all lower branches that turn yellow or brown, snapping them off flush with the stem. If you want to limit the growth of the plant to the height of the stake, pinch off the top when it reaches the end of the stake. But if a taller plant is desired, add a stake extension and let one of the suckers grow—suckers usually become part of the main stem and not separate stems when high up on the plant.

Even in single-stem pruning it is advisable not to prune much after the plants reach the top of the stakes. Instead, let foliage develop at the top to protect fruits below from sunscald.

Small shoots, or suckers, developing where the leaf stem joins the main stem should be removed in single-stem pruning.

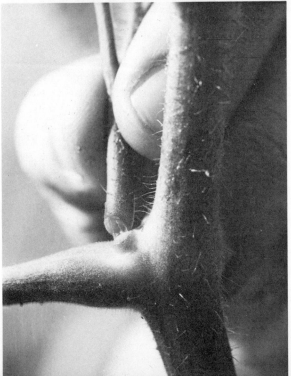

It is best to prune out small shoots by hand.

Tomatoes grown outside in the typical English fashion, each plant pruned to one stem and allowed to carry four trusses of fruit.

Double-stem Pruning. The object here is to train plants to *two* main stems, a method that will produce more fruit than single-stem pruning and gives more protection against sunscald. The technique is almost identical to single-stem pruning except that one sucker becomes the second main stem or stalk. The original stem or leader is stem number one and the sucker chosen for stem number two is usually the sucker that has developed just below the first cluster of tomato flowers. Both main stems are of course tied to the stake. Prune out all other suckers, and continue to do so as the plant grows.

Multiple-stem Pruning. Three or more main stems are allowed to develop in multiple-stem pruning, the procedure the same as for double-stem pruning above. Far more fruit is produced this way, and better-quality fruit, but it is difficult to tie the three or more stems to a single stake. You can solve this problem by using two stakes for each plant, or by pinching off branches as they appear.

Retaining Suckers When Pruning. Some gardening gurus claim a threefold increase in fruit production by allowing all suckers to develop, to a point. Here the suckers (again, the tiny side shoots that sprout between the main stem and leaf stem) are all allowed to grow until they form blossom clusters. Then they are pruned so that no more growth occurs on the suckers after the blossom cluster. The suckers are finally tied to the stake. This method is easiest to use on single-stem plants, but can be employed in conjunction with all the pruning methods above. The theory is that plants retaining suckers will yield fruit not only between every other leaf node, as is normally the case, but on each sucker as well. More traditional gardeners claim that fruit size is smaller using this method and that it results in plants running amok with wild growth. Certainly the technique is worth a try on a few plants.

Defoliation. A pruning procedure that often hastens ripening of tomatoes by five or six days, defoliation is practiced after green tomatoes have formed and reached their mature size. Contrary to popular belief, light has little effect on ripening tomatoes—putting green tomatoes in a bright window, for example, will not hasten ripening. But temperature *does* effect ripening—tomatoes don't ripen well at temperatures below 65° F. or over 85° F. Therefore, removing leaves from tomato plants and allowing more sunshine to reach the fruits will raise the temperature and hasten ripening. Fruits can also be spotted easier if some foliage is removed from plants, and the fruits will dry off faster after rains, reducing the development of mold and rot diseases. But defoliating plants does make them weaker, exposes them to possible damage by unexpected frosts, and causes an increase in anthracnose rot. (See Chapter Thirteen.)

COMPANION PLANTS FOR STAKED TOMATOES

If you stake your tomato plants and prune them (even if you just stake them well), you'll find that there's usually a lot of room between plants for raising other vegetables. Lettuce, radishes, beans, onions, beets, carrots, cabbages, peppers, and almost any vegetable that doesn't take up a great deal of space are the best choices, though even elbow-room vegetables like asparagus (which has an affinity for tomatoes) can be grown with tomato plants if the garden is planned right. Experiment with a few of these companion crops, but be sure to supply them with extra water-soluble fertilizer so that nutrients aren't stolen from the tomato plants. Some companion crops can be sown long before tomatoes are set out. I usually sow lettuce in the tomato rows early in the spring. When the tomatoes are set out, just enough lettuce is harvested to make room for planting holes and the rest is left growing between tomato plants.

Herbs also grow well alongside tomato plants. Basil might be set near them, for there is hardly a tomato dish that isn't improved by a bit of basil, "the tomato herb." Chervil, dill, marjoram, wild marjoram (oregano), sage, and tarragon are other herbs that are called for in many tomato recipes. Chives grow well with tomatoes, too, as does parsley. Rocket, another herb (sometimes called roquette and ruca), is especially good in a tomato salad as a substitute for lettuce—it provides all the seasoning that is needed for the salad save oil.

Some vegetables and herbs aren't compatible with tomatoes—fennel and kohlrabi are two. Others improve tomato quality greatly. Even the common weed stinging nettle aids the tomato. Tomatoes growing in the weed's vicinity resist spoiling, are sweeter, and have unusual lasting quality—probably because formic acid from the nettle roots is absorbed by the tomato plants near them. The subject of plant symbiosis, plants aiding each other, is fully discussed in Chapter Twelve, where a number of such plants are noted. Keep in mind, however, that tomato planting time is also the time to set out most helpful tomato companions.

Chapter Ten

FEEDING 'EM— INCLUDING COMPOSTING, SPECIAL FERTILIZING PROBLEMS, AND VITAMINS

Not long ago an enterprising gardener in Tavares, Florida, fertilized his front lawn with three tons of sludge from the local sewage plant. Instead of lush green grass, he ended up with some forty thousand tomato plants! "Well, I don't know how you can say it politely," he explained, "but the plants came from people. Tomato seeds aren't digested. They just pass through. Ripe or green, I like tomatoes, but I'm not a nut over them, and I'm sure I won't be after this is over."

The odds are perhaps more than forty thousand to one that you won't have similar troubles nurturing your tomato patch, even if you choose sludge as a fertilizer. Tomato feeding is a relatively simple matter. A tomato plant will yield (and often yield well) if it is just set in prefertilized ground and watered once a week. However, this isn't always the case, and tomato yield will invariably increase if plants are given more attention. The tips following will prove especially useful to anyone who wants to break tomato production records or raise really mammoth fruits. To take all guesswork out of fertilizing, however, soil in the tomato patch should first be tested as recommended in Chapter Three.

Prefertilize the ground in which tomatoes are to be set and feed transplants with a starter solution as described earlier. *Don't fertilize again until the first fruits appear on the plants*—in fact, wait until the fruits are as big as half dollars for best results. This is important, for too much nitrogen fertilizer in the first stage of growth is one reason why plants don't set fruit—the plants aren't encouraged to change gears from the vegetative stage to the fruiting stage of growth. However, when small golf-ball-sized fruits do develop on the plants,

TOMATO-FEEDING BASICS

plenty of nitrogen is needed. There are a number of ways to supply this. *Don't use all of these methods, only one:*

1) Feed plants with any of the many liquid tomato fertilizers on the market. These are applied as per directions on the container with a watering can, sprayer, or hose feeder. To make a gallon of a less expensive liquid fertilizer from dry 5-10-5 fertilizer, tie up four ounces of the 5-10-5 in a piece of cloth. Swish this around in a gallon of water for a few minutes, squeezing the cloth occasionally, and then leave the cloth in water for two days. Remove the cloth bag and dispose of it—the material remaining will only be the crushed rock used for fertilizer binder. Water plants every ten days with this solution, a gallon to each plant. Note that while the solution is much cheaper, it will not supply as many nutrients to your plants as a carefully formulated commercial preparation. The same applies to manure teas, which are made by soaking composted manure in water. (See instructions on any manure bag.)

2) Uniformly scatter a heaping teaspoon of 5-10-5 fertilizer around each plant and mix it into the top half of soil eight to ten inches from the stem. Repeat this procedure once or twice a month.

3) Dig in a side dressing of about a half pound of ammonium nitrate per fifty feet of row. Repeat the application two more times at three-week intervals.

4) Use one of the new single-application, controlled-release fertilizers that provide nutrients only as plants need them. As plants utilize the nutrients in a controlled-release fertilizer, its outer surface is gradually worn away—much as a child licks a lollipop. MagAmp is one of the newer single-application fertilizers. About ten pounds of it is banded along the sides of the plants at planting time. (See Chapter Eight for banding methods.)

USING ORGANIC
FERTILIZERS

In the long run organic fertilizers such as animal manures and leaves will improve soil in the tomato patch (the results can often be seen in as little as a year), whereas commercial fertilizers can actually harm soil by overuse over a period of time. Long-term overfeeding with commercial fertilizer leads to an accumulation of soluble salts in the soil that will cause the locking up of certain nutrients. Another disadvantage is that weeds thrive on inorganic fertilizers and, finally, organic matter in the soil is rapidly depleted by overfeeding with inorganics. One should always be careful to follow directions when applying commercial fertilizers. "Use half as much twice as often," is, in fact, a good time-honored rule to follow if you're ever in doubt.

The value of organic matter in the soil is unquestioned. It improves soil tilth, increases the water-holding capacity of soils, and through its decay, releases nitrogen and other nutrients for plant use. Carbon dioxide from decaying materials helps bring minerals into solution, making them available to plants. Organic matter also stimulates root production, maintains the mycorrhizal fungi that aids plants in the absorption of nutrients, and even reduces insects like nematodes by encouraging the growth of parasitic fungi. That is why farmers have plowed manure and cover crops into their fields for generations. Organic fertilizers do have certain drawbacks, though. Their chief disadvantages are their relatively high cost (though many, like leaves, are free for the taking) and their slow action. Organic fertilizers do not feed at all in cold early spring weather—plants cannot absorb organic matter until it has decayed in warm soil and broken down into simple chemical forms. On the other hand, organics are long-lasting, feeding plants over the entire growing season, while inorganics may run out by fall.

All and all, organic fertilizers are of much more value to the tomato gardener and they certainly benefit the ecology far more than the inorganics. Try to use them whenever possible—dig them into the soil in autumn, or incorporate them into planting holes when setting out tomatoes. They will never burn plants or leach away during a heavy rain as inorganics will. You'll find through experience that little additional fertilizing will be necessary other than, perhaps, a commercial starter solution in the spring, and an organic fish emulsion can even be used as a starter. Many books are devoted to organic fertilizers, government pamphlets abound on the subject and at least one magazine (*Organic Gardening*) is dedicated to their use. Make it a point to learn more about this important subject. In the meantime, to give you an idea of the values of just a few of the many common organic fertilizers, their percentages of nitrogen, phosphorus, and potassium (or potash) are listed below:

Fertilizer	*Nitrogen %*	*Phosphorus %*	*Potassium %*
Inorganic 5-10-5	5.0	10.0	5.0
Bone meal	0.0	20–25	0.0
Blood, dried	8.12	2.5	.5
Coffee grounds	2.0	0.4	0.7
Cottonseed meal	6.0	7.3	1–2
Cow manure (composted)	0.6	0.15	1.5
Eggshells	1.2	0.4	.15
Fish scraps	6–10	7.0	.8
Horse manure (composted)	0.7	0.25	0.55
Leaves (oak)	0.8	0.4	0.2
Pine needles	0.5	0.1	.03
Poultry manure (fresh)	4.5	1.5	1.5

Fertilizer	Nitrogen %	Phosphorus %	Potassium %
Rock phosphate	0.0	25–30	0.0
Seaweed	0.6	.10	1.3
Sewage sludge	5–10	3–13	0.0
Sheep manure (fresh)	2.5	1.5	1.5
Tea grounds	4.0	0.6	.40
Wood ashes	0.0	1–2	3–7

COMPOST FOR
TOMATOES

Composting is simply the disintegration process in which organic materials are broken down by the action of bacteria and fungi. When these materials are broken down in a compost pile, they decay more quickly than they would in the soil, yielding the dark rich, crumbly compost that has so many uses in the garden. Compost can be mixed in planting holes when tomatoes are set out, dug into the garden, or used as a side dressing for plants throughout the season. It can't burn plants like commercial fertilizer and benefits the soil in the same ways that organic materials do—only faster. A well-made compost has a fertilizer value of about 2-3-5.

There is really no mystery about composting. Almost any plant material can be used to make compost—leaves, grass clippings, vegetable matter, spoiled hay, even weeds can be used. Compost-making kits complete with safe chemicals to speed up decomposition are available at garden outlets. Many elaborate composting techniques exist (and you should consult a composting manual as you get deeper into the subject), but compost can easily be made in about two weeks by following the informal method described here. Do this throughout the gardening season and you'll have enough compost to cover over six thousand square feet:

1) Set aside an 8-by-4-foot area in the spring and start the compost pile by filling the area with a four-foot-high heap of leaves and grass clippings (or whatever other organic material you have on hand). If the pile is low in nitrogen (grass clippings, green plant matter, etc.), add manure to it.

2) Mix up the material in the pile and shred it all into small pieces with a rotary mower or shredder. This will speed up disintegration by exposing more surfaces to attack by bacteria and fungi.

3) Make four- to six-inch layers with the shredded material, covering each layer with about one inch of rich soil. Water each layer down and add another layer until the heap is four feet high.

4) By the third day the heap will have begun to heat up. Check it with a thermometer and add more nitrogen if it hasn't. Keep the heap moist and turn it with

a pitchfork or shovel on the fourth, seventh, tenth, and fourteenth day. By the end of two weeks the compost will be ready to use, although it will take on a richer more crumbly look if you let it decay a while longer.

Covering a compost pile securely with black polyethylene plastic not only eliminates odors and unsightly heaps but speeds up decomposition. Other organic materials you can use in the heap include cornstalks, corncobs, vines, sawdust, wood chips, slaughterhouse residues, pine needles, nutshells, bark, coffee grounds, and cocoa bean hulls. Some of these (like wood chips) will take longer to break down than others and need more nitrogen. If you use an acid organic material, such as oak leaves, sprinkle a little lime over them before shredding. It's estimated that the leaves from one large oak tree will make $15 worth of compost.

Tomato plants, like all living things, need many foods or nutrients. Air and water supply them with carbon, hydrogen, and oxygen. The remaining nutrients, which come from the soil or fertilizers, are, with the barest of explanations: **SPECIAL TOMATO FERTILIZING PROBLEMS**

Nitrogen—needed for green, healthy leaves and protoplasm.
Phosphorus—enables plants to mature crops.
Potassium or Potash—for stronger stems and good root systems.
Calcium—helps roots and other plant parts, sweetens overacid soil.
Sulfur—makes protoplasm.
Iron—for chlorophyll, the green coloring of leaves.
Magnesium—helps leaves manufacture starch and sugar.
Zinc—used by the plant's enzyme system and for leaf-making.
Copper—for the enzyme system and for helping plants grow to their proper height.
Boron—for steady plant growth.
Molybedenum—used by the plant's enzyme system.
Chlorine—necessary, but for unknown reasons.

Generally, all of the above elements, save the first three, will be in sufficient supply for tomato growing in any garden soil. That is why the first three elements —nitrogen, phosphorus, and potassium—are generally the only nutrients included in the so-called "balanced" fertilizers (5-10-5, etc.) available on the market. Feed as directed with a balanced fertilizer and you'll usually have no problem with tomatoes. *But*—and this is an important but—many soils are deficient in other nutrients and need to be corrected. That is why the soil tests recommended in Chapter Three are a good idea. Have one made, follow the recommendations for improving soil, if any, and you'll know exactly how often and with which nutrients you'll have to fertilize. You can, however, tell whether tomato plants are deficient in certain nutrients by observing the plants while they are growing. Here are the most common signs of nutrient deficiency along with some corrective measures:

Nitrogen Deficiency—very slow growth of tomato plants; leaves turning light-green beginning at the top of the plants; small, thin leaves with purplish veins; stems that are stunted in growth, turn brown and die; flower buds yellowing and dropping off; reduced yield. To correct feed with a fertilizer high in nitrogen, such as blood meal.

Phosphorus Deficiency—foliage turning purplish beginning with the undersides of leaves; small leaves; slender stems; slow growth; late fruiting. Correct by using bone meal at planting time, phosphate rock or any fertilizer high in phosphorus at other times.

Potassium Deficiency—slow, stunted growth; crinkled young leaves; yellow-green margin on older leaves along with bronze-colored spots; leaves turning brown and dying; hard woody stems that don't increase in diameter; brownish, underdeveloped roots; low yield; fruits with uneven ripening and lacking in solidity. Correct by adding wood ash, potash rock or other potassium-rich fertilizer to soil. Mulching also helps here. (See Chapter Eleven.)

Calcium Deficiency—yellow *upper* leaves; weak flabby plants; thick, woody stems; stems spotted with dead areas; short, brownish roots. Correct by using agricultural grade limestone at manufacturer's rates. Bone meal and super-phosphate are good sources of calcium, too. Crushed eggshells spread on the soil around plants—one shell to every four square feet—will also rejuvenate tomatoes deficient in calcium.

Magnesium Deficiency—brittle, curled-up leaves that turn yellow all over the plant, beginning with lower leaves (the yellow color deeper when farthest away from the leaf veins); late ripening; fruit lacking in flavor. Correct by using dolomite limestone or one tablespoon of Epsom salts mixed into the soil around each plant. Seaweed and seaweed extract are good sources of magnesium, too.

Boron Deficiency—black areas at the growing points of the stems; stunted stems; abnormally bushy plants; terminal shoots that wither and die; bees neglecting the flowers; fruit with dark or dry areas. Correct with manure or a little borax (one ounce dissolved in twenty-five gallons of water). Vetch or sweet clover dug into the soil will also help.

Zinc Deficiency—abnormally long, narrow leaves that may turn yellow and become mottled with dead areas. Occurs primarily in peat soils. Correct by using plenty of manure of any kind, or seaweed, shredded cornstalks, hickory, poplar, or peach leaves.

Iron Deficiency—spotted, pale areas on young leaves; yellow leaves on top of plants; death of new shoots, plant tissue. Correct with dried blood, manure, or sludge. Most weeds are rich in iron, too. Cut down on the use of lime in the garden.

Copper Deficiency—stunted growth of shoots; very poor root systems; curled, flabby bluish-green foliage; undersized plants; few or no flowers. Correct with ample amounts of manure, wood shavings, or bluegrass clippings.

Manganese Deficiency—slow, stunted growth; extremely light-green leaves turning to yellow (dead spots appear in the yellow and spread); few blossoms, little or no fruit. Correct by digging in manure, or grass clippings. Oak leaf mold and alfalfa are also good here, as is seaweed or seaweed extract.

VITAMINS FOR TOMATOES

Tomato plants may or may not need vitamins—scientific research in this area hasn't been completed, though it is fairly well established that all plant roots need vitamin B. Some growers, however, have found that vitamins used as a fertilizer adjunct make their plants bear better. If you want to experiment with vitamins and tomatoes—and be ahead of most everyone on the block—try SUPER-THRIVE, which is available from Vitamin Institute, 5409 Satsuma Avenue, North Hollywood, California 91603. SUPERTHRIVE contains over fifty hormones and vitamins. It is applied during weekly waterings at a rate of one drop per gallon of water. The manufacturers will provide complete directions.

GIBBERELLIC ACID

A number of gardeners have experimented on tomatoes with the growth-promoting stimulant, gibberellic acid, which is derived from a microscopic Japanese fungus. Gibberellic acid is thus far a laboratory tool and isn't recommended unless you have plenty of plants to experiment upon. The instructions on the package should be followed carefully if it is used. Gibrel is one trademarked product featuring gibberellic acid.

OVERFERTILIZING REMEDIES

If your plants have an overabundance of sparkling green leaves, exhibit abnormal green growth, and produce few blossoms (or blossoms keep dropping off), the plants may have received too much nitrogen fertilizer. To remedy this pinch off some of the excess leaves. Stop feeding the plants high nitrogen fertilizers and incorporate granite dust or bone meal into the soil. Next year make sure that too much nitrogen in the form of fertilizers, cover crops, or green matter isn't dug into the soil.

Chapter Eleven

LOVE AMONG THE LOVE APPLES:
WATERING, WEEDING, PROPAGATING, AND
PAMPERING TOMATOES

Only cucumbers and a few leaf vegetables have a higher water content than tomatoes, which are 93.5 per cent water (though a unique and very highly flavored water indeed). During the main growing season, tomatoes need about one inch of water a week from either rainfall or artificial means, and if the soil is very sandy, they will require nearly double that. Some market growers even give their plants two inches of water a week on the theory that tomato roots grow deeper than other vegetables.

Once tomato plants are set out, diverge from the "one inch of water a week" rule until fruit the size of half dollars has set on the plants. Water the same amount *but less frequently*—only every ten days or two weeks—in these prefruit stages of growth so that the vines are put under slight stress to develop tomatoes. Then go back to watering one inch a week. *Water regularly; do not wait until plants are wilting or flagging as some garden manuals recommend.* After fruit has set, it's very important to maintain an even soil moisture. Fluctuating dry and wet spells often bring on both stunting of tomato plants and blossom-end rot—the last occurring (for one reason) when plants growing rapidly with high soil moisture suddenly experience a hot dry spell. Irregular watering will also result in small fruits maturing slowly or not at all.

But remember—*never overwater*. Tomato plants with roots completely immersed in water will die of thirst, for water can move into their roots only in combination with oxygen. This is why good drainage is so important. The rule is to keep the soil uniformly moist, but not soggy or muddy. Keep in mind, too, that fruit cracking will occur if plants are watered heavily after a prolonged drought.

On the other hand, if tomato plants are only lightly sprinkled, their roots will grow up to the surface of the soil to obtain moisture—putting them in great danger of drying up and dying when the soil dries. The ground around the plants should be thoroughly soaked to a depth of six to eight inches when watering; approximately two thirds of a gallon of water per square foot of soil is needed to attain this saturation level. If you are using a sprinkler, set four or five coffee cans around the tomato patch. When these fill up with one to one and a half inches of water, the patch has been watered to a six- to eight-inch depth. Don't worry about tomato leaf roll after a thorough watering, either. No one knows exactly what causes this physiological reaction, but it is usually preceded by heavy watering or deep cultivating—the leaves rolling upward and toward the main stem. As new roots form on the plants, the leaf rolling will disappear.

Watering can be done at any time so long as tomato foliage isn't watered too late in the day. The old myth about midday watering being dangerous is just that —the *best* time to water is at midday when the sun is bright. Midday watering not only lets plant foliage dry in the sun but increases humidity around the plants and cools the leaves, decreasing transpiration and eliminating the midday wilting that checks growth. On a still evening, water may remain on leaves for several hours before drying. Anytime tomato leaves are wet for more than three hours, they are extremely susceptible to leaf-blighting or leaf-spotting fungi. Don't compound these problems, which occur naturally with long rains, by using the garden hose carelessly.

During the gardening season, water bills normally go up 50 to 100 per cent. Use good water conservation practices and you can practically eliminate this unnecessary expense. Water conservation not only cuts down on bills, but protects a precious natural resource, eliminates hot summer chores, and even benefits tomato plants. Except for mulching, discussed further on, the best watering moneysavers and work savers follow:

▪ The U. S. Department of Agriculture has found that seltzer used as a water substitute is more beneficial to many plants than plain H_2O. USDA scientists discovered that leaf lettuce treated with carbonated water produced three times more leaf in six weeks than lettuce sprayed with water, and that chrysanthemums so treated had mature blooms two weeks before untreated specimens. This trick might work with tomatoes as well.

▪ Sea water has also been used as a water substitute. Experiments indicate that numerous vegetables and fruits, including tomatoes (which proved mildly tolerant to the treatment), can be saved by irrigating with brackish or slightly salty water during a temporary emergency, *although its prolonged use can be detrimental.*

- The rain barrel, far from a modern water-saving device, came into its own again during the last great drought in the United States a decade ago. Rain barrels, painted to match either house trim or shingles, are best set under gutters to catch runoff water from the roof. The drain downspout is placed inside the barrel, which should be covered to discourage mosquitoes. The water collected can either be stored to ladle out later, or the barrel can be fitted snuggly with a pipe and tap at the bottom. The barrel is often rigged at the top with a pipe, "T" connection, and hose so that overflow from heavy rains will be siphoned off to nearby tomato plants.

- Installing a well on your property is another excellent way to get water for tomatoes. There are many books and pamphlets available on the subject, or you can consult any plumbing-supply dealer. Just be sure to have the water tested to determine its acidity or alkalinity, or any excess of mineral salts. The quality of well water may necessitate changes in fertilizing practices. When pH is high or low, for example, the availability of certain trace elements is reduced.

- You can even "manufacture" your own H_2O. A newly developed solar-still method, originated by the USDA, actually "makes" water. It is a simple matter to build a solar still using polyethylene plastic. Just dig a two-foot hole in the ground in an out-of-the-way area, place a bucket in its center and cover with plastic held in place by stones. The sun will draw water from the soil, which will collect in the bottom of the sagging plastic and drip into the bucket. In a week about 2½ gallons of pure water can be collected, more if a larger hole is dug.

- For most plants a little water is worse than none at all, so don't permit runoff in your tomato patch. Soak the ground when watering, even if you have to use cooking, wash, and shower water (small amounts of soap and detergent will do tomato plants no harm). Another method is to recess open-ended No. 10 cans in front of each plant and water into these. Or set inverted bottles of water near the plant roots so that watering is gradual and there is no waste.

- Still another simple way to water deeply and cheaply, is to tie an old sock around the garden hose nozzle and lay the nozzle next to a plant. Soaker hoses with small pinpoints in them are also good. A more efficient (and expensive) new variation on these are the deep irrigation systems now available that are patterned on a system developed in Israel. Modern deep irrigation both saves water and results in yield increases of 30–100 per cent in tomatoes by insuring that plants get exactly the amount of moisture they need at all times. Literature on the subject can be obtained from Watersaver Systems (Box 2037, Pomona, California 91766), or Chapin Watermatics (Box 298, Watertown, New York 13601). A garden watering kit designed on this principle that covers 625 square feet is available from DuPont Company (Wilmington, Delaware 19898) for $14.95.

■ Chapin Watermatics (above) offers a semipermanent Spray Stake system, where watering stakes are hammered into the garden hose near tomato plants. The stakes are adjustable for direction and amount of spray.

■ New "garden computers," electronic moisture-sensing controllers, monitor soil moisture by remote buried sensors. The sensors feed back information to the control unit, which completely automates watering by watering only when the soil needs it. Some models are under $100. Literature can be obtained from Agrotronics Manufacturing Company (Box 1248, Barstow, California 92311).

■ If water is hard to come by in your locality, try to use drought-resistant to-

*A twin-wall soaker hose with small perforations, extended the length of
a tomato row, gives plants ample water with very little runoff.*

*Spray stakes, attached
to a hose, water
tomatoes efficiently and
economically.*

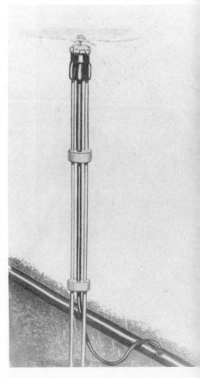

mato varieties. (See Appendixes I and II.) And remember that the most important conservation measure is to keep the soil in good tilth. Let it shed no water and aerate thoroughly. As noted in Chapter Three, improving the soil is a relatively easy matter. Cover crops such as inexpensive annual grass seed can be planted and turned under to provide organic matter. Moisture-holding humus like compost, peat moss, manure, and well-rotted leaves can be introduced to enrich the earth and make it more friable. Do so and many of the preceding measures may not be necessary.

TOMATO-PATCH
WEED CONTROL

While it's true that weeds are only uncultivated plants that grow where they are unwanted or whose virtues have yet to be discovered, they do cause estimated losses of over four billion dollars on farms alone in the United States every year. Weeds remain with us despite the billions of dollars spent annually to destroy them, and weeds will always be with us, if only because of their obstinacy—weed seeds buried in a bottle on the Michigan State University campus in 1890 were found to be viable in 1970, eighty years later! So the villains must be dealt with. Weeds definitely compete with tomatoes, as they do with all vegetables, for moisture, nutrients, and light. They also harbor insects and diseases. Chickweed, for example, can host the virus known as spotted wilt which is so dangerous to tomato plants.

Tomatoes just cannot compete with weeds and probably more tomato plants do not produce their maximum yield because of them than any other reason. Ironically, the better your soil, the more weeds you'll have, but they can easily be handled with a little effort. Besides mulching (below), there are three basic ways to beat weeds:

Cultivating. In cultivating keep the area around tomato plants free of weeds for at least four feet in all directions. Remove the weeds by hand (after a rain is a good time), or cut them off shallowly (about one inch below the soil) with a sharp hoe, weed knife, or cultivator so as not to disturb tomato plant roots. Wait until the weeds are one inch tall to do so and use the weeds in the compost heap so long as they contain no seeds. While cultivating with a hoe or other implement, also loosen up the soil. Cultivating has as its object not only weed control, but the introduction of oxygen from air into the soil and the prevention of erosion and leaching of water. Loosening the soil to a depth of one inch will accomplish these objects.

Herbicides. Herbicides aren't recommended here, for they aren't essential to control weeds in the tomato patch; the alternatives of cultivating or mulching (see the following section) are just as effective, consume about the same amount of time, and are far safer for you and the environment. But since this is supposed to be a complete tomato guide, I must report that many home gardeners control weeds with herbicides. Amiben, or DCPA, is a favorite and is applied to the soil immediately after setting plants out. The plants won't be injured if you follow the manufacturer's directions. A cheaper alternative is suggested by the USDA. "Carefully directed sprays of full-strength Stoddard solvent cleaning fluid will kill established weeds between mulched rows without damage to the tomato plants," the agency advises. "Use this spray when there is no wind. Use a low pressure that gives a coarse spray. Thoroughly wet the weeds."

Soil Fumigation. Reported but not recommended here for the same reasons as herbicides. In this method the soil is fumigated with methyl bromide before planting. The ground has to be spaded up and worked smooth with a rake. Then a plastic sheet is placed over the area to be treated. The edges of the sheet are covered with soil to keep the methyl bromide from escaping and the gas is released (in the quantity recommended by the manufacturer) under the plastic sheet. Finally, the sheet is kept in place for forty-eight hours, when it is removed and the soil is cultivated for aeration. Tomatoes may be planted seventy-two hours after this—in a total of five days altogether. Fumigation will control practically all weeds and nematodes and many diseases in the garden, but the manufacturer's precautions for the use of methyl bromide should be followed very carefully.

Just one additional important warning concerning weed killers. If you are using selective weed killers (chemicals that kill some plants without harming others) *anywhere* on your property, be especially careful. For example, even minute traces of 2,4D, which is used to kill broad-leaved weeds on lawns without harming the grass, can cause growth reactions and even death to tomatoes and other plants nearby.

SAVE WORK BY MULCHING TOMATOES

By covering the soil around tomato plants with a mulch (a layer of leaves, straw, or most any organic or inorganic material), you will not only increase their yield but reap many other benefits. Mulches reduce the need for watering and weeding by as much as 95 per cent, protect roots of plants against temperature extremes and sudden changes, and improve the soil and feed plants if organic mulches are used. They also protect the soil against compaction by traffic, prevent soil erosion on hills, encourage feeder roots in the rich upper layers of soil, prevent plants from being splashed with disease spores in heavy rains, reduce pollution caused by soil-applied pesticides, produce vital carbon dioxide as they decay, and can even stimulate growth in plants by reflecting light to them. Add to this long list of benefits earlier maturity of fruits, larger tomatoes, cleaner, less damaged tomatoes, and easier harvesting and it's hard to resist mulching plants in the tomato patch.

Mulching has been nature's way of protecting and providing for plants since time immemorial, the moist, rich, spongy carpet of the forest floor a perfect example of mulching. The word mulch itself is an ancient one. In Anglo-Saxon *melsc* was "mellow," and in an old German dialect molsch meant "rotten"; these words led to the English mulch, which was first used in the mid-seventeenth century. Today mulches are divided into organic materials like leaves and hay, or inorganic ones like black plastic or aluminum foil. The organic mulches are preferred because they add nutrients to the soil, whereas the inorganic ones control weeds better.

A grass mulch around tomato plants controls weeds and conserves moisture, helping plants to bear better.

Mulches have several minor disadvantages. They do contribute to the holding of excessive soil moisture on poorly drained soils in wet seasons, which harms tomato root systems. They can also attract slugs and mice and conceal mole burrows. But these are usually problems that can easily be solved. Most bad results from mulching are due to improper applications. Mulch properly once and you'll never garden any other way.

To begin with, *don't* mulch early in the spring. Application of organic mulch at this time will slow up the natural warming of the soil. As an insulating blanket,

Tomato plants growing on black plastic mulch in which holes have been cut to receive plants. Black plastic mulch conserves moisture, controls weeds, warms the soil, and hastens maturity of plants.

organic mulches reduce solar radiation into the soil and as a result, frost hazards are greater. So wait until the soil has thoroughly warmed up before you apply mulch. If you keep a mulched garden from year to year, just push aside the mulch and plant the tomatoes, pulling the mulch back in place when summer weather arrives.

Before mulching tomatoes when the ground is warm, clean out all weeds, cultivate the ground an inch deep, and water it thoroughly. Then pile organic or inorganic materials that are insect-, disease-, and weed-free around the plants to a depth of about four inches. Do not place the mulch closer than a few inches from the plant stems. If you still have trouble with slugs and other pests attracted to the mulch and eating your plants, try the corrective measures noted in Chapter Twelve. As for the materials that can be used for mulches, these number in the hundreds. Forty or so are listed below along with their advantages, drawbacks, and any special ways to apply them. *Apply all of these four inches deep except where noted:*

Aluminum Foil—repels some insects, and reflected light from it often increases yields. Weigh down with stones around plants or cover with a heavier, more attractive mulch.

Asphalt Paper—forms a long-lasting barrier against weeds, and can be covered with a more attractive organic mulch.

Bagasse (Chopped Sugar Cane)—long lasting, clean, light-colored, and holds water well. Apply two inches deep. Sometimes sold as chicken litter.

Bark, Shredded—lasts a long time, is aesthetically pleasing, adds much humus to the soil, retains moisture, and will not blow away. Apply two inches thick.

Black Plastic—see *Plastic Film*

Buckwheat Hulls—dark, attractive, long-lasting, and do not mat—but blow away when dry. Apply three inches thick.

Cloth—burlap, old rags, old rugs, etc., can be laid between tomato rows if appearance is not important.

Cocoa-bean Hulls—tends to pack and mold; has a chocolate odor for a few weeks. Best mixed with other mulches. Apply two inches thick.

Coconut Fiber—hard to get, but is long lasting and attractive.

Coffee Grounds—an excellent soil conditioner with attractive color, but is slightly acid. Sprinkle a little lime on them.

Compost—use when only half rotted; cover with other mulches.

Corncobs, Ground—need a little nitrogenous fertilizer sprinkled over them; may attract vermin. Apply two inches thick.

Corn Stalks—can either be shredded or used whole covered with more attractive mulches.

Cottonseed Hulls—good, but hard to obtain.

Dust Mulch—this simply means shallow cultivation of the soil to create a layer of dust that prevents upward movement of water and thus reduces evaporation. Some experts say all it really does is kill weeds by the act of cultivating.

Excelsior—a good mulch, long lasting, non-packing, and weed free, but highly flammable. Apply two inches thick.

Fiberglass Matting—repels insects, is permeable to air and water. A brand called Weed Chek is widely available.

Glass Wool—good but tends to blow away unless covered with chicken wire or other mulch. Apply two inches thick.

Grass Clippings—tests show this is one of the best mulches to repel nematodes on tomato roots and increase yield; however, grass clippings will mat, ferment, and smell if used fresh, and can harbor insects. Use *dry* grass clippings three inches thick or mixed with other mulches; if you use the dry clippings alone, try adding a half cup of blood meal per bushel.

Gravel, Marble, and Quartz Chips—good on their own or for holding down other mulches, and protect well against mice. Experiments at the Colorado Agricultural Experiment Station showed that tomatoes mulched with black gravel yielded 10.27 tons per acre, while white-gravel-mulched tomatoes yielded 8.8 tons, and unmulched plants yielded 2.86 tons. Black gravel-grown plants were also freer of blossom-end rot. The black gravel mulch was applied 1½ inches deep.

Green Manures—cover crops, usually legumes, grown in place and cut for mulch.

Hay—excellent but may need fastening down. Don't use hay going to seed. Apply six inches thick.

Leaf Mold—rotted leaves, especially nutritious for soil.

Leaves—oak leaves are the best; avoid types like maple that tend to pack unless they are shredded first. You can dust oak leaves with a little lime if you like, but tests show their acidity doesn't effect the pH of the soil when used as a mulch. Wet down and apply one to two inches thick.

Licorice Root—attractive, non-packing and long lasting, but flammable and hard to obtain.

Mushroom Soil, Spent—has good color, benefits garden soil.

Newspapers—the newsprint repels insects and the wood pulp fertilizes plants, but newspapers can be unattractive and can form a tight mat so that air doesn't reach plant roots. Use four to six thicknesses around plants and cover with a more decorative organic mulch that will also help break the paper down. Newspaper ashes can be used as a mulch, too.

Peanut Hulls—excellent, but can blow away and may attract vermin.

Peat Moss—dries out and crusts too easily. Must be kept wet or it won't permit water to reach plants and will even suck water from the soil below to meet its own needs. Peat moss is best incorporated into the soil, but if it is used as a mulch, choose chunky types and keep them stirred up and wet. Not recommended for areas receiving little rainfall if rain is the only source of water.

Pecan Hulls—good mulch but hard to obtain.

Perlite—blows away too easily, must be weighed down with another mulch. Yields no nutrients.

Pine Needles—attractive and useful, but flammable.

Plastic Film (Polyethylene)—unattractive but very effective. Black (or green) plastic is preferred because it doesn't permit weed growth like clear plastic. Black plastic also absorbs the sun's heat during the day more than organic mulches do and radiates the heat back faster at night; thus plants mulched with it are less liable to frost injury than those mulched with organic materials. Tomato plants generally yield more when mulched with black plastic; its chief disadvantage is that it doesn't improve the soil any. Transplant tomatoes into holes cut through the plastic *after* it is put down as a mulch. The procedure is simple. On a windless day soak the ground thoroughly. Mark off the area to be covered by the plastic, dig furrows four inches deep along the edges of this space, unroll the plastic, and anchor it in the furrows with soil. Then make holes in the plastic and set in your transplants. Perforate the plastic for aeration. Remove the sheet at the end of the growing season and use it again next year if it's in good condition. Black or green plastic is best in an .0015 thickness and is available from most garden supply stores.

Salt-marsh Hay—one of the most effective mulches because it contains no weed seed, doesn't mat down, and is light and airy. Can be used for many years as it doesn't break down quickly.

Sawdust and Wood Shavings—sawdust is a good mulch that is not toxic to plants as is often written; however, it does consume a lot of nitrogen in decomposing, depriving plants of this nutrient. Apply three inches thick and add either a half cup of blood meal or a cupful of nitrate soda per bushel. Wood shavings should be treated the same way.

Seaweed—if you live along the coast, this is a superb mineral-rich, growth-promoting mulch. Can be placed directly around plants or composted first.

Straws—particularly good are wheat and oat straws, which are coarser than hay but more durable. Straws, however, are more flammable than hay and contain weed seed. Apply four inches thick.

Rocks and Stones—attractive, warm up the soil, and add trace elements to the soil as they imperceptibly wear away. Ring each tomato plant with them a half inch from the stalk and five inches outward, piling the rocks three inches high. Round, flat rocks weighing a few pounds are the easiest to work with, but many rocks and stones can be used. "Stones" in a wide variety of shapes can even be made from cement.

Tobacco Stems—don't use on tomatoes, as they cause mosaic disease.

Vegetable Peelings—good, generally, but don't use old tomatoes or potato peelings or plants—tomatoes used as a mulch can cause canker and bacterial spot; potatoes can introduce verticillium organisms into the tomato patch. Compost these waste products instead so that heat will destroy any diseases. One nurseryman reports that *composted* tomato vines from the previous year used as a mulch greatly increases tomato yield.

Vermiculite—see Perlite

Water—in recent USDA experiments, 6-ml. clear plastic bags filled with water increased tomato yields by 20 per cent. By absorbing intense heat from the summer sun at midday and reducing heat loss at night, a water mulch allows you to plant earlier in the spring and extend your planting into the late fall. Heated water in the bags warms up the soil even more.

Weeds—if weeds haven't gone to seed, they can be used as a mulch. So that they don't root again after a heavy rain, put them atop four or five layers of newspaper, which is a good way to get rid of two "waste products" at the same time.

Wood chips—more attractive than sawdust or shavings, but need the same amount of nitrogen.

If you have the space for more plants in the garden, there are two easy ways to propagate tomato plants from old ones. These tricks are especially valuable for areas with long growing seasons, but can also be used to propagate plants that will be brought into the house for the winter (see Chapter Fifteen):

1) Layering. You'll need unpruned bushy plants with lots of stems. Take a twelve-inch-long stem and bend it carefully to the ground, covering all but the top four inches with soil or mulch. Protect the tops from cutworms by using cardboard collars, and keep the soil wet. In about two weeks the stem will have formed roots and can be cut from the parent plant and placed elsewhere. By using larger stems or covering less of the stems with soil, you can propagate bigger plants.

PROPAGATING
TOMATOES

2) Cuttings. Just cut off four-inch lengths from side branches with a sharp knife. Dip them in rooting hormone and stick them in sand or even the garden soil. They will root and can be transplanted to their permanent location in two weeks.

POLLINATING 'EM

Sometimes tomato blossoms aren't pollinated in still weather. The tomato blossom is what is called a perfect flower, with both male and female parts, and is wind-pollinated: wind-blown pollen from the anthers is trapped on the curly stigma and works its way down the pistil tube to fertilize the ovum. To encourage pollination during still, windless periods, gently shake the plants. Some gardeners even shake the flowers with an electric toothbrush or other electric vibrators.

PAMPERING 'EM

It's said that W. C. Fields cussed out his tomato plants occasionally, but recent evidence suggests that you'd be better off treating yours with tender loving care—speak gently and carry a big watering can might be the motto. You can gently apply all of the early-bird tricks noted in Chapter Seven to hasten tomato ripening at any time of the season. Or better yet you might be able to *talk* your plants into ripening faster or yielding more and bigger fruits. It isn't my intention to go into plant sensitivity theories here; there are enough books available on that subject. But the fact remains that many gardeners have attributed bumper tomato crops to treating their plants like people. It may well be that these gardeners are simply more aware of and responsive to their plants' needs as they talk or sing to them and are rewarded accordingly. Or tomatoes may really be a lot like people. Anyway, here are some of the specific stratagems that have been suggested:

▪ Some gardeners play music to their tomato plants, claiming that this stimulates growth. Scientists say that sonic stimulation may be a valid horticultural technique, and various experiments seem to have indicated this, though they are far from conclusive. Classical music, it's said, works better than rock. All of Bach, "Rhapsody in Blue," "Nights in the Gardens of Spain," and Ravi Shankar's classical Indian sitar music have been highly recommended. So if you hang a transistor radio in the garden, don't leave it on an acid-rock station; if you do, the plants will "cringe, lean sharply away from the sound, and die in a few weeks," advises one professor. Try the recording *Music to Grow Plants By* (Pickwick International Records) if this theory appeals to you.

▪ Tomato plants are also responsive to the human voice, say other experimenters. Don't be a W. C. Fields, they warn. Never scold tomatoes; speak gently at all times. Content, too, is quite important here. Don't even try to curse a tomato gently! He/she will surely know. Encourage your plants, spur them on, tell them things will get better when it's too cold, wet, when the insects are coming, etc. At least one scientist sings "Happy Days Are Here Again" to her tomato friends.

- Handle tomato plants very gently, as well, we are advised. They respond well to petting and tender care. A noted psychiatrist claims that a person with a "true green thumb" can heal wounds in plants by simply touching them. Another university researcher has reported that even "water that a green thumb touches will make plants grow at express speed." How do you get a "true green thumb," though, unless you're born with one? "Brown thumbs" only make matters worse.

- Don't dare think bad thoughts when close by tomato plants if you subscribe to the plant sensitivity theory. Several polygraph researchers have written that plants registered violent reactions on their machines when people even thought about harming them.

- When students studying sensitivity in plants talked about sex, the plants they tended showed signs of sudden excitation on their polygraphs. Which led them to believe that ancient fertility rites, in which humans made love in freshly seeded fields, may have been authentic stimulants to plant growth. You'd have to be careful of the seed bed to do that, but at least you can talk lustily to plants. Pick a prudish plant, though, and you might get a ripe tomato right in the mouth.

- A professor of natural science, at Blake College in California, found in a two-year research program into the "emotional lives" of various plant species that tomatoes are among the vegetables "most susceptible to kindness and flattery." So, if you do talk to them at all, lay it on with a garden trowel.

- Finally, two Scottish gardeners have created a veritable vest-pocket Eden on a half acre of soil that is mostly sand and gravel. They have accomplished this, they say, solely by "communion with the living plants" and "happiness and joy" in what they are doing. These gentlemen claim that all fruits and vegetables take pleasure when their exquisite flavor is appreciated by a human being and do their best when their efforts are most happily acknowledged. So take a salt shaker into the garden once in a while, eat a warm, sun-ripened tomato fresh off the vine, and praise its glories with your mouth full.

Chapter Twelve

THE WORMS AT THE HEART OF THE ROSE: CONTROLLING TOMATO INSECTS AND PESTS

People aren't the only animals with a penchant for tomatoes. An assortment of villainous vegetarians ranging from the slimy slug to the dragonlike tomato hornworm consider the plant and/or its fruits a great delicacy. Probably more insects feed upon tomatoes than any other cultivated crop, but, luckily, these pests can easily be controlled. We'll launch a direct assault on the most dangerous insect pests shortly, but first there are some general rules that should be observed.

RAISE HEALTHY PLANTS IN A CLEAN GARDEN

Strong, healthy, well-cared-for tomato plants are the least vulnerable to insect damage. Keep the tomato patch neat, clean, and free of weeds, for debris and weeds harbor destructive insects. Turn up the ground in the fall, too, to expose insects to birds and weather. Rotating crops in the garden is also important—plant tomatoes in the same spot year after year, for example, and nematodes will build up in the soil.

It's a good idea to closely examine any tomato plant that doesn't look healthy; you'll soon develop a sharp eye for harmful insects. Visit the garden once a day to check plants and if you notice morning damage to plants without finding the culprit, return to the tomato patch after dark with a flashlight. Use a magnifying glass of at least 5x enlargement power to examine any insects you find, placing them in a clear plastic bag and flattening the bag to hold them still. Harmful insects can then be hand-picked and dropped in a tin of kerosene or salt water.

Any plant that dies in the garden should be dug up carefully so that the roots can be examined for signs of insect damage or disease. And don't compost diseased plants that you pull out—burn them or carefully dispose of them in another way. Such plants often become a source of infection for next year.

It goes without saying that no diseased tomato plant that cannot be saved should be kept in the garden. Some gardeners keep badly diseased plants in place to show off as curiosities. No matter how carefully such plants are "quarantined," they can spread diseases to healthy tomato plants nearby.

Check the variety lists in Appendixes I and II, which note a great number of resistant varieties and tomato plants resistant to microscopic insects like nematodes. If you've had problems growing tomatoes because of diseases, select a variety from the lists for a type resistant to whatever tomato diseases may be prevalent in your section of the country. No tomato variety is resistant to *all* diseases, but hundreds of varieties have some immunity against one or more diseases.

Scientists estimate that 99 per cent of the 1.6 million known insects species are beneficial to plants. Learn which insects are helpful to tomato plants and don't harm them; in fact, introduce them into the garden where possible. The same applies to birds and other wildlife. An incomplete list of these "good guys" follows. It can be supplemented by a complete book on the subject like C. L. Metcalf and W. P. Flint's *Destructive and Useful Insects—Their Habits and Control* (McGraw-Hill):

Assassin Bugs. Long-legged brown insects ¾ inch long, with wings folded together over the body. They do bite people if bothered, but if left alone devour a good many harmful insects, grabbing them by the front legs and "assassinating" them.

Assassin Flies. Hairy black and gray flies about 1½ inches long with yellowish legs. They prey on many insects.

Birds. Birds can be a problem (see the Pest List farther on), but their good qualities usually far outnumber their bad habits. Little birds especially are apt to be meat-eaters and won't harm tomatoes. Most birds feed voraciously on slugs, aphids, and other harmful bugs.

Centipedes. Flat, brown, many-legged creatures with one pair of legs per segment (unlike the slightly harmful millipede, with two pair per segment), which often live under rocks. They dine on several insects, including slugs and snails.

Damsel Bugs. These resemble the assassin bug but are only about half their size. They feed on aphids and other small, soft-bodied insects.

Doodlebug. A plump brown insect ½ inch long with jaws like forceps. Doodle-bugs destroy ants, which often carry aphids. They dig a cone-shaped hole in the ground, hide at the bottom, and wait until an ant falls in.

Dragonflies. Large (2 inches long) insects with big eyes and transparent wings. Fast fliers, dragonflies catch other insects with their legs while airborne.

Flower Flies. Small (⅜ inch long) insects that resemble bees. Flower flies hover over flowers and pounce on aphids and other harmful insects.

Lacewing Flies. Delicate, ½ inch-long pale-green insects with golden eyes and filmy wings twice the size of their bodies. Lacewing fly larvae are about ⅓ inch long, yellowish and torpedo-shaped, with hairs on the body and jaws shaped like forceps. These larvae are called aphid lions because of the prodigious amount of aphids they eat. Lacewing larvae are also valuable because they devour hard-to-kill mites and other insects.

Ladybugs. Ladybugs, the best-known beneficial insects, comprise some three hundred fifty species around the world. They feed on aphids, mites, white flies, scale insects, and the eggs of other insects.

Moles. Moles can be destructive in the tomato patch (see the Pest List following), but they do eat grubs, cutworms, and other harmful insects.

Praying Mantis. Another familiar insect, often kept by kids as "pets," the praying mantis lives entirely on insects like aphids. Huge by insect standards (they range from 1½–5 inches long); they are greenish or brown and have triangular heads that they turn from side to side. Their front legs are held in a "praying" position, ready to seize other insects.

Spiders. Spiders destroy numerous garden pests. They have eight legs, insects six —count the legs if you're in doubt.

Toads. Keep a pet toad in the garden and it will devour many insects; including cutworms, slugs, and stink bugs. Toads can be attracted by keeping a clay pan filled with water in a shady part of the garden.

Trichogramma Wasps. Tiny microscopic insects whose black eggs are often found in the garden. The wasps deposit these eggs in the eggs of more than two hundred harmful insects. When the trichogramma eggs hatch, the larvae kill the embryos of their host eggs. So leave those black eggs alone.

Hornets, ground beetles, and many other insects can be helpful in the tomato patch—only the most common benefactors could be mentioned above. If you want to introduce beneficial insects to the garden, there are quite a few insectaries that sell ladybugs, trichogramma wasps, and other good guys. A pint of ladybugs contains about ten thousand of the little insects and costs only three or four dollars (try the Ladybug Beetle & Sierra Bug Company, which is really located on Lady Bug Lane in the town of Rough and Ready,

California 95975). A widely available product called Bug Bait will attract ladybugs, green lacewings, and flower flies to the garden. This method of insect control does work; many farmers, nurserymen, and public parks across the country are using it today.

Companion and trap plants combat troublesome insects when planted near tomatoes. These helpful plants either repel insects from the general area, or attract insects to them so that they can be trapped and destroyed in considerable numbers. Examples of trap plants are larkspur and white geraniums, which can be used to attract Japanese beetles away from tomatoes—when the beetles congregate on these plants, they are collected and destroyed. Companion plants, plants that *repel* insects from the tomato patch, include marigolds, which cut down on nematodes in the soil; begonias, which are never approached by aphids; and garlic, which repels many insects. Basil and other herbs also help tomatoes when planted nearby, as do nasturtiums, calendula (pot marigold), chives, parsley, cosmos, and coreopsis. There are some plants that you should *not* put near tomatoes, such as roses, grapes, or potatoes. All of these beneficial plants and their opposites are discussed under the specific insects that they fight or aid in the Pest List at chapter's end. Experiment with other plants, though. Observe combinations that seem to work in the garden. An excellent book on this subject is *The Organic Way to Plant Protection,* from Organic Gardening, Emmaus, Pennsylvania 18049.

There are numerous broad-spectrum pesticides on the market today that control a host of tomato insects and are far less dangerous than DDT—in fact, so many of these chemical tomato sprays and dusts are readily available that it isn't necessary to describe them at length. Needless to say, the manufacturer's directions for using these pesticides should be followed to the letter.

Less widely known, and recommended here, are the organic or biological insecticides and repellents. Some can be purchased at garden centers, others you'll have to make yourself, and all work safely and efficiently. Here is a list of a score or so:

Bacillus thuringiensis. Available under the brand names Dipel, Biotrol XK, and Thuricide, this product is very effective against cutworms and tomato hornworms. It uses fungi, bacteria, and viruses to stomach-poison the pests, but is harmless to humans.

Basic H. A commercial preparation made from soybeans that is deadly to aphids, thrips, and red spiders.

Compost Solution. A spadeful of compost soaked in a bucket of water for an hour or so and sprinkled over plants will both fertilize them and repel a number of insects.

Doom. A commercial bacterial insecticide that destroys Japanese beetle grubs and other insects by transmitting milky spore disease to them.

Entocons. A commercial bacterial insecticide that controls aphids, among other insects, by disrupting their growth or reproductive processes.

Garlic-Pepper-Soap Spray. Blend together four crushed cloves of garlic, four tablespoons of hot pepper, a cake of strong soap, and a cup of hot water. Dissolve in two gallons of hot water (the size of most watering cans), cool, and use for a general insect spray.

Garlic Spray. Press a garlic clove and mix the oily juice with a gallon of water for an all-purpose spray.

Ground Hot Pepper. Sprinkle on tomato plants for protection against several insects.

Hot-pepper Spray. Grind hot pepper pods and mix with an equal amount of water and a little soap powder for use against tomato worms.

Onion Spray. Chop onions finely in an electric blender or by hand and mix one tablespoon with a pint of water for an all-purpose spray.

Pickled Peppers. A pint (or peck) of pickled peppers put through a blender and sprinkled over plants makes a good general insecticide.

Pyrethrum. See *Rotenone*.

Rhubarb Leaves. Soak three pounds of the leaves in three quarts of water for an hour and sprinkle over plants infected with aphids.

Rotenone. A commercial preparation that is derived from tropical plants. It is of low toxicity to humans and kills many types of insects as both a stomach poison and contact poison. Rotenone works slowly, lasting about a week. A similar product, Pyrethrum, works quickly in a day or so. The two are often mixed together for best results.

Scent-off. A commercial preparation that repels dogs and cats.

Soapy Water. A soapy solution sprayed or painted on tomato plants serves as a good general-purpose insect repellent. Green soap or soaps made with fish or coconut oils are best, but any will do. Wash off the soap with water from a hose a few minutes after applying or it might harm plants. Some gardeners mix one cup of green soap with two cloves of garlic and three gallons of water for a soap solution.

Squell. A commercial product that repels squirrels.

Tomato-leaf Spray. Tomato leaves themselves contain solanine, an alkaloid that is a repellent to many insects, including aphids. Boil the leaves and stems in water and spray the solution on plants when it cools.

Vegetable Spray. Grind finely and mix two hot peppers, a large onion, a garlic clove, and a teaspoon of detergent and let this set in a little water for a day or so. Strain it and add the liquid to a pint of water for an all-purpose insect spray. You can use geranium leaves, mint, and other strong-flavored plants in the mix, too. Experiment.

Water. A fine hose spray of water will kill aphids and many other insects, or knock them off the plants. The water also dilutes the juices aphids feed on and turns them off so that they do little damage.

Wildcat Scent. A commercially prepared repellent that emits the scent of mountain lion, which scares off deer and other wild creatures.

Zip. A commercial product that is a taste repellent used as a bait for deer. Penco Therain works similarly.

Gardening Without Poisons by Beatrice Trum Hunter (Houghton Mifflin) gives more home remedies. Other specific organic repellents are mentioned as cures in the Pest List following.

Below is a "most wanted list" containing the thirty or so most troublesome tomato pests and ways to combat them. Wildlife, as noted, isn't much of a threat— wild animals generally prefer other vegetables to tomatoes. Neither will you be troubled by more than a few of these insect pests, unless you are especially unlucky, though most are generally distributed throughout the country. Don't panic if you notice a chewed leaf or other damage on one of your plants. Most likely the plant will survive and produce admirably, especially if steps are taken to help it. If a plant is dead or beyond help, remove it from the garden and burn it or dispose of it in another sanitary way.

Aphids. Aphids are small (less than ⅛ inch), round, soft-bodied plant lice ranging from pale pink to deep green in color. Clusters of them occur on tender stems and undersides of tomato leaves, sucking out plant juices and causing leaves to curl. Ants often spread aphids (which they keep as "milk cows" for the sweet secretion they exhude) from plant to plant. To control aphids organically, you can: 1) see that soil in the tomato patch is rich in humus, which these insects avoid; 2) plant aphid-repelling nasturtiums near tomato plants; 3) plant companion crops of garlic and chives close by tomatoes; 4) plant rhubarb nearby; 5) mulch tomatoes with aluminum foil (the aphids will be repelled by the sun's reflection on the foil); 6) prune off and dispose of tomato leaves with masses of aphids so that they will breed no more; 7) wash the aphids off tomato plants with soapy water; 8) control the ants that carry aphids by sprinkling bone meal around their holes.

Chemical controls for aphids include dusting them with vegetable dust, or spraying with Malathion. Spraying with safe insecticides like Rotenone and Pyrethrum is also effective. Some growers simply spray aphids with water once a day; this dilutes the juices the aphids feed on, encouraging them to leave for greener pastures. Others use a solution made by boiling tomato stems and leaves in water. Basic H, the biological spray made from soybeans, also destroys aphids.

Birds. A transistor radio left on in the garden often scares birds off—they appreciate neither classical nor rock music. Various commercial noisemakers and timed fireworks have also been successful, as have recordings of bird distress calls. Pieces of rope placed around the garden are sometimes mistaken for snakes by birds, white string wrapped around bushes seems like spiderwebs to them and broken mirrors or aluminum pie plates hung from plants scare them off, too. The trouble with all these controls—as with the time-honored scarecrow, which usually becomes little more than a roost for the feathered creatures—is that the birds soon become accustomed to them. Volch oil sprayed on tomatoes might keep birds away from ripe tomatoes—this harmless spray does protect other berries from birds. But birds are not usually much trouble in the tomato patch; they eat many insects harmful to tomatoes. It is crops like corn and peas that they ravage.

Blister Beetles. Medium-size (½ to 1 inch) gray, red, black, or striped beetles that feed on tomato leaves. The best way to combat these busy beasties is to pick them off plants by hand and crush them. But be sure to wear gloves, for as their name implies they can blister skin with a secretion called cantharadine (which, incidentally, is the chief ingredient in the deadly "Spanish fly" or Cantharides sold illegally as a so-called aphrodisiac).

Corn Borers. Corn borers are little (one inch) pinkish grubs that are the larvae of a brown moth and only occasionally eat tomato leaves and fruit. The best control against them is to crush the masses of corn borer eggs laid on the undersides of leaves. Rotenone spray is also somewhat effective. Don't plant tomatoes near corn if possible.

Cutworms. Smooth gray, black, brownish, or greenish caterpillars (up to 1¼ inch long) usually found curled up just under the soil. Cutworms cut off young plants evenly at the surface of the ground (uneven cuts indicate rabbit damage) and are generally a problem early in the season. There are a number of cutworm controls: 1) place stiff three-inch cardboard collars around plants at ground level (an inch into the soil and two inches above it) so that cutworms can't get to plant stems; 2) scatter mothballs or blood meal around plants to repel cutworms; 3) dig up the ground in early spring to expose cutworms, thus freezing or starving them; 4) keep the garden clean so that adult cutworms can't lay the eggs that turn into the destructive wormlike larvae; 5) line boards on dampened soil to lure cutworms underneath them where you can gather and destroy them. The biological insecticide Bacillus thuringiensis also controls cutworms. Chemical controls include dusting the soil with Chlordane after planting.

Deer. Wire fencing eight feet high or more is often the only protection against marauding deer, though a dog will also provide some control. Lining the garden with clothes drenched with perspiration—to warn of man's presence—has been suggested by some gardeners, as have mothballs and dried blood mixed together. Several commercial taste and scent repellents for use against deer are available, including one that emits the scent of a mountain lion.

Dogs and Cats. There are a number of products on the market that repel dogs and cats. They also dislike hot pepper sprays.

Earwigs. Fairly large (one–two inches) brown or black insects with or without wings that have a pincerlike tail at the end of the abdomen. Earwigs damage leaves on young tomato plants. They like to gather under boards and can be trapped and destroyed by laying boards on the soil. Encourage wild birds to nest near your garden—they like earwigs for breakfast—and, if you raise poultry, let your chickens or ducks dine on them in the garden. Sea sand spread around plants repels earwigs, as does mothballs, or tea leaves mixed equally with soil and peat moss. Commercial earwig bait used as per the manufacturer's directions also kills the pests.

Flea Beetles. Small (less than ⅛ inch) brown to black jumping beetles which make small round holes in young tomato leaves and whose larvae are small white worms which feed on the roots of plants. Flea beetles can be controlled by keeping weeds down in the garden, so that they can't nest in them and proliferate, and by planting basil near the tomatoes, which repels the beetles. Dusting with Rotenone or Pyrethrum also helps. So does sprinkling the plants with a soap solution. Flea beetles are so quick that they're practically impossible to catch by hand. They usually leave tomatoes early in the season, before they can do any irreparable harm.

Gophers. Try the same controls as for *Moles.*

Japanese Beetles. The adults are a shiny green and bronze, a half inch long, and have brown wings. Japanese beetles feed on tomato foliage. Try hand-picking or spraying with Rotenone to control them. If they are a real problem, don't plant tomatoes near roses or grapes and make sure no knotweed is nearby—these plants are all favorites of Japanese beetles. Odorless marigolds, white geraniums, and larkspur are flowers that seem to attract the beetles away from tomatoes; you can use them as trap plants and collect the beetles from them. The most effective control is the bacterial insecticide Doom, which gives Japanese beetle grubs (the small larvae that live in the soil) a fatal illness, milky spore disease, and breaks their breeding cycle.

Leafhoppers. Small (up to ⅕ inch) slender, pale green to brown, wedge-shaped, soft-bodied insects which suck juice from the undersides of tomato leaves and tomato stems. You'll notice leafhoppers by the dozens or hundreds if they're present. Curled tomato leaves with dark tips are evidence of their ravaging. Leafhoppers are particularly bothersome west of the Rockies, where they are called Beet Leafhoppers. Spray with Pyrethrum, Rotenone, or Malathion. Or dust with sulfur. Don't plant tomatoes near beets.

Mice. Mice are less of a problem if you use a gravel-like mulch instead of a grassy one. Mothballs repel them, as does garlic planted in the garden, and traps and poisons can be set out for them.

Mites (Tomato Erinose). Very small mites that cause tomato leaves to develop a white fuzzy look. Tomato erinose are usually a problem in the Southeast and on the Pacific Coast. A jet of plain water directed at the mites is one control, spraying with Rotenone another.

Moles. Mothballs and garlic planted in the garden repel moles, and castor oil sprinkled around plants helps, too. Inspect under mulch for mole tunnels and either set specialized mole traps for them or use anti-mole chemicals. Another solution is to plant tomatoes in a board-enclosed foot-high raised bed, lining the bottom of the bed with half-inch mesh wire. The mole plant (*Euphorbia lathyris*) makes an excellent mole repellent, but this South African species is difficult to obtain. Castor bean plants repel moles, too, but beans from this plant are poisonous if eaten.

Nematodes. Microscopic "eel worms," invisible to the naked eye, that are especially troublesome to tomatoes in the South, causing stunted plants and leaving swellings (galls) on plant roots that contain their eggs. Nematodes are usually controlled by rotating crops (never growing tomatoes or related crops like potatoes and eggplant in the same space for more than two years so that nematodes can't build up in the soil); or by planting one of the many nematode-resistant tomato varieties. (See Appendixes I and II.) Another control is to keep the soil well supplied with humus by digging in organic matter or by using a repellent mulch (generally grass, decayed leaves, or

The female golden nematode ends her life as a tiny cyst containing 200–500 eggs. The cysts eventually drop off. They may remain in the soil as long as eight years and release nematode larvae every season.

Root-knot disease is caused by nematodes which hatch and enter the roots, where they feed. Their feeding causes enlargements, called root-knots or galls, up to three-quarters of an inch in diameter. These galls make tomato plants susceptible to other diseases, such as fusarium wilt, and increase the severity of such diseases.

water hyacinths), both of which will encourage the growth of fungi that attack nematodes. Marigolds (especially African marigolds) planted near tomato plants will exhude a substance from their roots that will repel nematodes a year after planting.

In a recent experiment the USDA found that ordinary sugar kills nematodes by drying them out. Five pounds of sugar used per one hundred pounds of soil killed all nematodes within twenty-four hours. Still another organic control is to place slices of the wild or mock cucumber (*Echinocystis lobate*) on the ground around tomato plants. This bitter fruit, often called the wild balsam apple, seems to repel nematodes. Or you can pour asparagus juice (the water asparagus is cooked in) over the soil. Chemical nematode controls include sterilizing the soil with Nemagon or fumigating the soil. (See Chapter Eleven.)

Pinworms. Small (up to ¼ inch) gray to green, brown-headed, slender worms which tunnel into tomato stems and fruit and web leaves together. Pinworms are most common in the South and California. Control by hand-picking whenever possible. Also try garlic spray and the botanical spray Rotenone.

Potato Bugs or Beetles. The adults are yellow and black, striped, hard-shelled beetles about ⅜ inch long. Immature potato bugs are soft, reddish, humpbacked forms about a half inch long. Both adults and young feed on tomato leaves. Hand-picking is a good control as is crushing any eggs found on leaves. Dusting with Rotenone also works. Potato bugs are one of many reasons for not planting tomatoes near potatoes.

Potato Psyllids. Little (less than ⅛ inch) jumping plant lice, tan to brown in color. Potato psyllids feed on foliage, causing the leaves to roll and become yellow or purple. They also stunt plants and deform fruit. They are general over the United States but most severe on the eastern slope of the Rocky Mountains. Remove infested foliage by hand where possible, or use the spraying controls recommended for aphids. Again, don't plant tomatoes near potatoes.

Rabbits. Blood meal mixed with a gallon of water and sprayed around the garden will repel rabbits. So will Epsom salt sprays (though too much of this will harm the soil) and mothballs. Tying the family dog up in the garden area is also recommended. "Alive" or "have a heart" traps are used by some gardeners, who trap the rabbits unharmed and transport them to a wilderness area. Other gardeners say a chunk of liver soaked in a bucket of hot water thirty minutes makes a good rabbit-repellent spray. Still others bury bottles filled with water around the garden, only their necks sticking out of the ground. (This probably works because of lights reflected off the glass or wind whistling over the bottles.)

Raccoons. If they are a real problem you can try hanging a transistor radio in your garden—raccoons dislike loud noises. Lights left on in the garden also scare them off and the same "have a heart" traps used for rabbits are good here, too. Raccoons are sometimes discouraged by garlic cloves left around plants and cayenne pepper sprayed on plants.

Red Spiders. (See *Mites.*)

Slugs. Dark, slimy, soft-bodied, snail-like forms up to four inches long that are voracious night feeders. Slugs feed on tomato foliage and even on fruit near the ground. They hide under rocks, boards, mulch, and other objects in the daytime and travel at night, lubricating their nocturnal paths with a slimy mucouslike secretion. It takes them about eight days to travel a mile, even when they swing from plant to plant via mucous ropes. Carefully inspect the garden, especially under rocks, mulch, etc., and destroy any slugs you find by dropping them into kerosene. You can also put boards in the garden to attract them, or use cabbage leaves for the same purpose. As slugs need lubrication to travel, placing a circle of dry sand, sea sand, ashes, sawdust, or hydrated lime around plants will stymie them. Keeping plants staked also helps, as do coarse scratchy mulches like hay. Replacing soil at the base of tomato plants with rocks helps to prevent slugs from crawling up plants.

TO KILL SLUGS: 1) use commercial methaldehyde baits placed at the base of plants; 2) place shallow pans of stale beer near plants that the slugs will crawl into and drown in (beer attracts thirty times as many slugs as commercial baits, says the USDA); 3) use a pan of one tablespoon of flour and ⅛ teaspoon of yeast mixed with a cup of water in the same way as beer; 4) use solutions of grape wine, blackberry wine, or vinegar in the way you would use beer; 5) sprinkle salt on the slugs or spray them with two tablespoons of salt dissolved in a quart of hot water (though excess salt residue in the soil is harmful); 6) try new products like Snail Snare and Slug-Geta, which dehydrate and kill the pests rather than poison them. In the early spring you can help destroy the slug population by turning over the soil so that slug eggs are exposed and dry out and die in the sun. Try not to have the soil too acid also—slugs thrive on an acid soil.

Snails. Same controls as for slugs above.

Squirrels. Squirrels sometimes eat ripe tomatoes off the vine. Mothballs and commercial preparations very similar to them repel squirrels to a degree. So does red pepper. Also use the "have a heart" traps mentioned under *Rabbits.*

Vaseline coated on tomato stakes will keep squirrels from climbing them if that is a problem.

Stalk Borers. Medium-size (up to one inch), yellowish, spotted worms that bore into tomato stalks just above the roots and are especially injurious to tomatoes in the southern United States. Stalk borers can be controlled by dusting with Rotenone spray or Sevin. Hand-picking of eggs on undersides of tomato leaves is also worth a try.

Stink Bugs. These are not a major problem with tomatoes. There are probably more than five thousand species belonging to the stink bug family. They are variously marked with green, orange, white, brown, or black, and have a shield shape and an unpleasant smell. Stink bugs are juice suckers that attack many vegetables. The best control is to keep the garden clean of weeds, in which they congregate, and hand-pick them from plants.

Symphylids. Small white worms that live on the soil and cause tomato plants to turn yellow, stunting their growth. Dusting with the pesticide Sevin controls them.

Tomato Fruitworms. Large (up to two inches), greasy, pale yellow to dark green or gray worms with darker stripes running lengthwise over their backs. Fruitworms eat into tomato fruits, moving from tomato to tomato. They are also called corn earworms or ballworms and are particularly destructive in Cali-

Tomato fruitworm larvae devastating a ripe tomato.

fornia and the South. Organic controls are hand-picking, and spraying with Rotenone or garlic sprays. Where fruitworms are a problem, don't plant tomatoes near corn. Dusting with Sevin is the chemical control.

Tomato Hornworms. The tomato hornworm is a giant (up to four inches) green, white-barred worm with a harmless thornlike horn projecting from the rear —probably the biggest and ugliest bug in the tomato patch. Hornworms adults are large hawk moths. Tomato hornworms are ravenous eaters of tomato fruits and leaves. They're easy to spot on close inspection of plants, which will have nearly skeletonized leaves as a result of hornworm damage. Hornworms usually deposit their eggs on the underside of leaves. Organic controls include hand-picking and destroying them by dropping them into kerosene, treating tomato foliage with hot pepper spray, and using trap plants like dill on whose foliage they are more easily spotted. Basil or borage planted next to tomatoes will repel hornworms, as will the organic spray Bacillus thuringiensis. *Don't* hand-pick any hornworm with white eggs attached along its back. These are the eggs of trichogramma wasps, a parasite that will hatch out, feed on its host, and then attack other hornworms.

White Flies. White fly larvae are small ($\frac{1}{16}$ inch) flat, greenish forms with white waxy spines and feed on undersides of leaves. Adults are small, white, mothlike flies which flutter out in clouds when disturbed, and thus are popularly called "flying dandruff." They are mostly a problem for greenhouse and indoor plants. White flies feed on the leaves of tomato plants. If they are a problem in the garden, try planting mint or tansy around tomato plants to repel them. Garlic sprays are also effective. In the greenhouse the parasite *Encarsis formosa* is often introduced to feed on white flies and is very effective.

Woodchucks. Are repelled by blood meal like rabbits (see above), or can be trapped.

The most formidable of tomato pests—Protoparce quinquomaculatus, or the tomato hornworm.

Chapter Thirteen

BATTLING THE BLIGHTS AND OTHER
TOMATO DISEASES

Catfacing . . . ghost spot . . . rot . . . canker . . . wilt . . . mosaic . . . cloudy spot . . . curly top . . . virus . . . blight . . . puffiness . . . mold. . . . The tomato is surely susceptible to more exotic diseases than any other vegetable grown in the home garden, but many gardeners have raised tomatoes for years and never encountered one of the scores of troubles that can befall them. If you observe the basic rules laid down for protection against insects—proper cultivation, clean gardening practices, etc.—you shouldn't have many troubles with diseases. Other general precautions are to be sure to use treated commercial seed if you plant tomatoes from seed, or to treat seed saved from the garden yourself (see Chapter Sixteen); not to irrigate tomato plants from above, especially at night, as this spreads disease; and to stay out of the tomato patch when foliage is wet.

If your locality is especially subject to one or two specific diseases, there are many disease-resistant varieties that can be planted, though no tomato is resistant to *all* diseases. Check Appendixes I and II for the best resistant types. If trouble persists, the following list of the most serious tomato troubles offers some specific remedies:

Anthracnose. This rot disease attacks ripe tomatoes when plants are grown in infertile, poorly drained soil. Anthracnose lesions are small, circular, slightly sunken, water-soaked spots that became darker and more depressed or develop ring markings. Controls: Enrich soil and improve drainage; rotate crops from year to year; clean up debris in garden; protect against rain splash by mulching properly; dispose of all diseased fruits so that they don't spread the disease. There are no resistant varieties for anthracnose. Chemical control is dusting with Zineb.

Bacterial Canker. A disease caused by contaminated seed-bed soil and infected seeds. The first symptoms are wilting, browning, and rolling of leaves on the

The new hybrid Super Fantastic produces large fruits on a plant with excellent disease tolerance.

The beefsteak-type Beefmaster yields super-tomatoes in the two-pound range and is a good disease-resistant type.

side of plant. Then the fruit becomes infected and shows "birds-eye" spots—whitish-brown spots with a white halo around them. Controls: Use only treated seed, or treat seed yourself (see Chapter Sixteen); buy certified canker-free plants; rotate crops, as the disease can live in the soil a full year; keep the garden clean; especially don't dig tomatoes or potatoes, or their foliage into the garden—they harbor the bacterial canker.

Jet Star, which does well in the North, produces giant fruits and has high disease resistance.

Homestead, a tomato for the South with good disease tolerance.

Burpee's VF Hybrid tomato, a bicentennial-year introduction, has excellent resistance to wilt diseases.

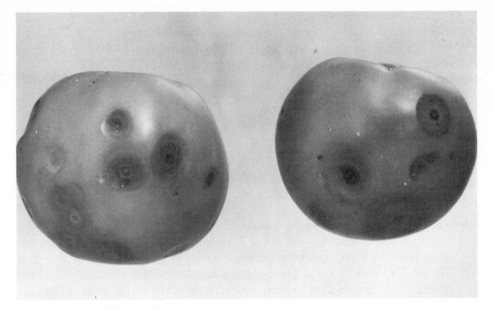

Tomatoes showing symptoms of anthracnose. Spots are sunken and have targetlike markings.

Bacterial Spot. Bacterial spot is common in warm rainy seasons. Lesions are small, dark, slightly raised dots (often with a water-soaked border) that can reach a quarter inch in size. Wind-blown rain spreads the disease. Controls: Same as for *Bacterial Canker* above.

Bacterial Wilt. Often called brown rot, this disease rapidly kills the entire plant, without spotting or yellowing of leaves. Stems at ground level will reveal water-soaked pith and a slimy ooze that exhudes from the stem when it is

Tomatoes showing early (left) and late (right) stages of bacterial spot. The spots are raised, but the centers are sunken and have a rough, scabby, appearance.

pressed (which distinguishes bacterial wilt from fusarium and verticillium wilts). Controls: Don't grow tomatoes or any related crops in the area for four to five years; grow seedlings in pasteurized soil; carefully dispose of all stricken plants; use disease-resistant varieties like Venus.

Blossom Drop. A tomato trouble rather than a disease, blossom drop most often occurs in hot, dry periods, and during unexpected periods of cold, rainy weather. Large-fruited varieties are especially susceptible. Controls: Keep plants mulched or well irrigated; do not feed plants large amounts of nitrogen, which encourages green growth at the expense of fruiting; grow resistant varieties in hot, dry climates like the Southwest; shake the plants to encourage fruit set in the blossoms; use blossom-set hormones. Other diseases listed here cause blossom drop, too, but they are always accompanied by other symptoms. Hotset and Porter are good resistant types.

Blossom-end Rot. A physiological disease caused by moisture fluctuations and a lack of calcium uptake from the soil to fruit during dry weather. The first symptom is a slight water-soaked area near the blossom end of the fruit. The sunken leathery lesion turns black and enlarges in a constantly widening circle, finally ranging from a mere speck to fully half of the tomato. This rot often appears on the first fruits of tomato plants set unusually early in cold soils and usually appears after a long dry spell. Controls: Keep soil moisture constant by mulching or careful irrigation; make sure the ground is well drained to allow proper root development; cultivate no closer than one foot from plants to avoid root pruning; avoid giving plants heavy dosages of nitrogen, which aggravates calcium deficiency in the soil (use superphosphate fertilizer instead); have soil tested for calcium deficiency, excess salts, and excess alkalinity, all causative factors, and ask your state agricultural exten-

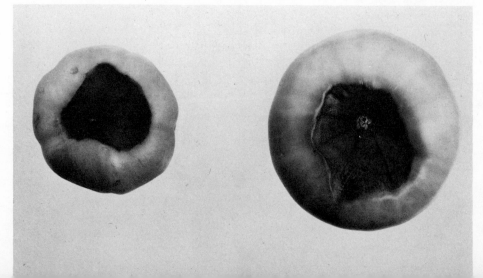

Tomatoes showing symptoms of blossom-end rot. Spot is dark, sunken, and leathery.

sion service for corrective recommendations. Staked plants are less affected by blossom-end rot than unstaked ones. There are several resistant varieties, including Rutgers and Marglobe.

Buckeye Rot. Water-soaked brown or grayish-green spots near the blossom end of the fruit where it touches the soil. This fungus disease only affects tomato fruits. Controls: Make sure the ground is well drained; take care in watering or mulching plants so that they won't be splashed by soil; stake plants so that fruits will not touch the soil; rotate crops from year to year.

Catfacing. Badly malformed and scarred fruit (appearing to some like a cat's face) at the blossom end of tomatoes, with bands of scar tissue, swollen bulges, and puckering of fruits. Extremes of heat and cold, drought, or pesticide injury (especially from the common insecticide 2,4-D), all cause abnormal development of the tomato flower, which results in catfacing. Hybrid varieties are not as susceptible as standards.

Chlorosis. Yellow tomato foliage, especially between leaf veins, that is due to excess lime or not enough iron. Controls: Have a soil-acidity test made and add iron sulphate if the soil is too alkaline.

Cloudy Spot. A blemish of both green and ripe fruits that is initiated by the feeding punctures of stink bugs. On green fruits the blemish is characterized by cream-colored ⅛- to ¼-inch spots, which change to yellowish spots on ripening fruits. Stink bugs can be controlled by the garlic or vegetable sprays given in Chapter Twelve.

Collar Rot. Young transplants are affected with this disease at the surface of the soil. Control: Dip transplants in a Meneb solution before planting.

Cracking of Fruits. Occurs after heavy rains or heavy irrigation, especially when there is high humidity. The growth cracks can be circular or radial and can become infected with a kind of rot. Controls: Remove tomatoes from the vine and use when they first crack; plant crack-resistant varieties like Campbell's 17, Heinz 1350 and Marion.

Cucumber Mosaic. A virus that causes the formation of misshapen, shoestring-type leaves and severe stunting in young plants. It is spread by aphids that carry it from nearby weeds, flowers, and vegetables. Controls: Keep the garden weed-free; control aphids (see Chapter Twelve); don't plant tomatoes near cucumbers, melons, or peppers, which can be infected and serve as a source of the virus; keep the virus host plants phlox, petunias, hollyhock, and zinnias away from tomatoes.

Curly Top. Leaves on the plants curl and twist upward. Leafhoppers spread this virus disease and should be controlled. (See Chapter Twelve.) Other controls include setting plants closer together than usual (from six to nine inches apart), and not planting beets, the leafhopper's favorite food, near tomato plants. Shading of plants with muslin tent structures is also practiced with some success, and late variety tomatoes aren't as susceptible as early ones. Chemical control is spraying with Malathon. Payette is a resistant type.

Damping-off. A disease caused by various fungi that results in indoor seedlings flopping over in their flats. See Chapter Five for controls.

Double-virus Streak. A combination of several mosaic viruses. Dead areas form along the veins in the leaves of the plants, and brown streaks appear on the petioles and main stems. Dry shrunken spots that resemble speckling with tomato juice form on the green fruits. Control is the same as for *Tobacco Mosaic* farther on.

Early Blight. Primarily a foliage blight caused by a fungus, but also causes fruit rot around tomato fruit stems in late fall. Early blight is characterized by brown irregular spots with concentric rings in a target pattern on the older, lower leaves. These spots soon enlarge to up to half an inch in diameter, run

Tomato leaves showing symptoms of early blight. Note targetlike markings.

A tomato affected with early blight.

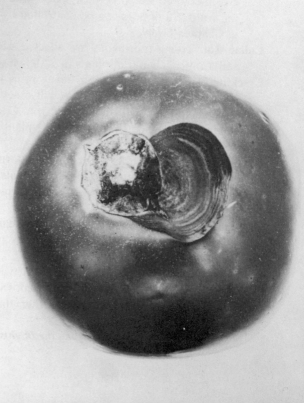

together, and cause the leaves to turn brown and usually drop off—often all the leaves drop off the lower half of the plant. Spotting and girdling of the stems also takes place. Robbing a plant of its leaves decreases the size of the fruits and exposes them to sunscald, and when the tomatoes themselves are attacked by early blight, they are useless. Controls: Keep the garden clean, as early blight fungus overwinters in old tomato debris and weeds; buy commercial seed, which is unlikely to be infected with the fungus; don't plant seeds close together in the seed bed if you use seed saved from the garden—this will spread the disease rapidly if it occurs; use varieties like Floradel or Manalucie, which are somewhat resistant. Chemical control is spraying with Maneb fungicide.

Fusarium Wilt. A common disease caused by a fungus that can live in the soil many years but only causes infection when soil temperature reaches 75–85° F. Symptoms are yellowing of the older lower leaves, which soon die, other leaves following until even the top portions are yellow and wilted and there is an overall wilting of the plant. Infection occurs through the root hairs, goes into the vascular system, and blocks the flow of water and nutrients into the plant. A brown discolored streak forms about ⅛ inch under the

Tomato plant showing symptoms of fusarium wilt.

surface of the main stem and usually extends to the top of the plant; slice the stem near the soil line to locate this symptom. If you have had fusarium wilt in the garden, be sure to rotate your tomato plants; don't plant them in the same place more than once in four years. Other controls include buying seedlings grown in clean soil, using sterilized soil to grow your own seedlings, and keeping the garden clean of weeds and refuse. Greenhouse soil must be pasteurized if the disease strikes once. By far the best precaution against fusarium

Tomato leaf with gray-leaf spot. Spots are small and irregularly shaped, and have a grayish-brown glazed appearance.

Tomato with advanced symptoms of gray mold.

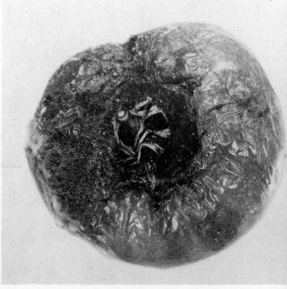

wilt is to buy one of the many resistant varieties available. (See Appendixes I and II.)

Ghost Spot. Characterized by small white circles surrounding a green center on the fruit surface, this fungal disease usually appears on the shoulders of small green tomatoes. Fruits generally develop to normal size and shape, are good to eat, but blemished. Controls: If the disease is prevalent in your area, do not plant under cool, damp conditions; don't plant tomatoes in any area of the garden where ghost spot has occurred.

Gray-leaf Spot. Small grayish-brown spots on the undersides of leaves characterize this fungal disease, which occurs mainly in the Southeast. As the disease progresses, the leaves yellow and drop, resulting in low fruit yield. If you have had gray-leaf spot on tomatoes, be sure to remove and destroy all plants in the garden at the end of the season, for the fungus is carried over on the remains of discarded plants. The best control is to use any of the varieties resistant to gray-leaf spot such as Marion, Manalucie, Floradel, and Tropic.

Gray-mold Rot. A fungal disease that attacks tomato fruits, leaving grayish-green to grayish-brown decayed spots that penetrate deep into the flesh. The disease begins with gray spots on tomato leaves and black markings on the stems. Mostly a greenhouse disease (though prevalent in the fields in Florida), gray-mold rot can be controlled by lowering the humidity and raising the temperature so that it won't spread.

Graywall. The cause of this disease, which results in blotchy ripening of fruit with internal browning, isn't fully understood. Experts suggest that shading of fruit has a lot to do with the disease and that nitrogen fertilizers should be avoided so as not to stimulate heavy foliage. Try not to plant in shady areas, too. Weeds, especially plantain, host a virus that may cause graywall, and should be eliminated from the garden. There are a number of resistant varieties, including Tropic-Red and Manalucie.

Hard Core. There is no effective control for this trouble, which is caused by temperature fluctuations, especially low night temperatures, and results in tomatoes with hard centers that are unpalatable. Early varieties are more susceptible than others.

Internal Browning. See *Graywall*.

Late Blight. A fungus disease that often occurs during long periods of muggy weather with cool nights, and strikes only once every ten to twelve years. Greasy black areas appear on the leaf margins, soon consuming the entire leaf, and a fine gray mold can be seen on the underside of the leaf during wet periods. Both green and ripe fruits become corky brown on their surfaces and take on a texture resembling an orange peel. The rot remains firm but makes the fruit inedible. Controls: Don't plant tomatoes near potatoes, which can become blighted and spread the disease; don't compost either infected potatoes or tomato plants; buy certified seedlings free of blight; look for first signs of disease (gray-green spots that become brown and hard) and burn infected plants so that they cannot spread the disease. Try disease-resistant varieties like New Yorker or Surecrop. Chemical control is Bordeau spray or Zineb dust application.

Leaf Mold. A fungus disease whose symptoms are yellowish or greenish spots, followed by purple mold growth on leaves. Usually occurs at its worst in damp, rainy weather. Try not to water plants at night to control leaf mold. In the greenhouse, keep humidity low, air circulation good, and temperature at 60° F. so that moisture won't collect on leaves. There are a number of resistant varieties, including Manalucie, Manapal, and Vetomold.

Leaf Roll and Curl. A physiological disease that commonly occurs in wet spring

Tomato leaves showing the large dark blotches characteristic of late blight.

Green tomato affected with late blight. Note the infected tissue.

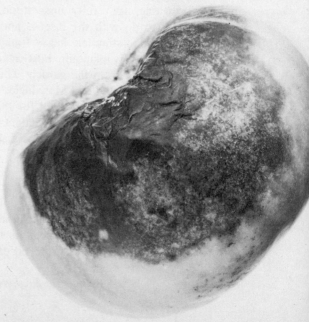

weather, especially on plants set in poorly drained soil. Some early varieties have genetic leaf curl and close cultivation or pruning also causes the disease. Leaf roll is really nothing much to worry about, for as temperatures rise and soils dry out, the symptoms disappear and normal growth resumes. No damage is done to fruits that develop later. Plants can lose a lot of their leaves, though, with loss in fruit *quality*. Controls include planting on well-drained soil, mulching with straw, not staking or pruning tomatoes, and cultivating shallowly.

Leaf Wilting. See *Fusarium Wilt*.

Nailhead Spot. Small tan to brown spots on fruit. See *Early Blight*, for which causes and controls are similar. The variety Marglobe is virtually immune to nailhead spot.

Psyllid Yellows. See *Potato Psyllid*, Chapter Twelve.

Potato Y Virus. A serious disease which results in yellowing of younger leaves, drooping plants, and purplish streaks on stems. It is carried by aphids (see controls in Chapter Twelve) and commonly occurs in tomatoes planted near potatoes.

Puffiness. Puffiness, or pockets, is a physiological trouble of tomatoes that results in the fruits becoming hollow, soft, and light in weight. When puffy fruits are cut, large air pockets are found in the cavities normally occupied by seed-bearing tissues. No decay or discoloration follows as a result of puffiness, but the fruits are of poor, unusable quality. Since unbalanced nutritional conditions created by heavy applications of nitrogen favor the development of puffiness, the use of superphosphate or bone meal together with only *moderate* amounts of nitrogen is a good control.

Purple or Bluish Leaves. This symptom usually indicates a phosphorus deficiency, especially when coupled with spindly plants. (See Chapter Ten for corrective measures for this and other nutritional deficiencies.)

Rots (minor). There are many minor tomato fruit rots, most of them of importance only to large-scale growers and shippers. These rots are often a problem if fruits have punctures in them. If you store any tomatoes green, check for small holes in them before storing. If they have any, use the tomatoes at once. (See Chapter Eighteen for green tomato recipes.)

Septoria Leaf Spot. A fungus disease that attacks plant foliage, blemishing it with gray-centered, water-soaked spots and eventually causing almost all foliage to fall off. Septoria leaf spot is most common during wet weather. The best control is to keep the garden free of refuse and decayed plants, which spread the disease.

Soil Rot. This disease is spread by a soil-inhabiting fungus, the symptoms small,

Tomato leaves with symptoms of septoria leaf spot. Spots have dark borders and light centers with dark specks.

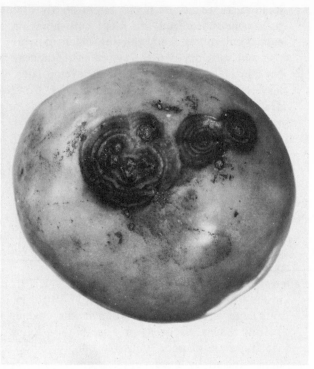

Tomato showing symptoms of soil rot. Note broken tissue and narrow targetlike markings.

brown circular spots on the lower halves of the fruits. The circles disappear as the fruits ripen, the spots become less evident, and the decayed area enlarges. The skin over the affected area is often split, unlike the similar *Buckeye Rot*. Soil rot is spread when tomatoes touch the soil or are splashed by rain. Controls are to stake tomatoes well above the ground, use a thick straw mulch around plants, and avoid planting tomatoes in poorly drained soil.

Spotted Wilt. A virus disease found in both garden and greenhouse tomatoes. There is a bronzing of the leaves caused by many tiny dead spots on the young top leaves; the tips of the stems show dark streaks; and ripe fruits have numerous large, raised spots consisting of alternate concentric rings of yellow and red. The spotted wilt virus (which isn't soil borne) is transmitted only by thrips from infected flowers, vegetables, and weeds. If the disease is a problem, keep tomato plants away from dahlias, calla lilies, nasturtiums, petunias, zinnias, onions, lettuce, celery, spinach, peppers, potatoes, mallows and jimsonweed, all of which harbor the thrips that carry the virus. Disease-resistant types include Pearl Harbor and Anahee.

Sunscald. Caused by fruits being overexposed to intense sun due to severe pruning, sparse foliage on a plant, or loss of foliage from diseases. Very common in the Southeast. Fruits are blemished with light-gray scalded spots that often become infected with fungi. Control by not pruning plants with light foliage, or by using heavy-foliage varieties. If plants do become badly defoliated, cover any fruits on them with straw.

Tobacco- or Tomato-Mosaic Virus. The most common virus disease of tomatoes, tobacco mosaic causes plant leaves to become mottled with yellow and green spots and to develop a rough texture. The mottled areas often turn brown and die. Infected plants are usually stunted and bear few fruits. Tobacco-

Tomato leaf showing mottling (yellow) caused by tobacco mosaic.

*Tomato plant affected
with verticillium wilt.*

*Tomato showing
downward curling that
is characteristic of
2,4-D weed-killer
injury.*

mosaic virus is carried by aphids and other sucking insects and overwinters in the roots of ground cherry, horse nettle, jimsonweed, nightshade, bittersweet, matrimony vine, plantain, catnip, Jerusalem cherry, and other related plants. Controls are to keep all weeds down around plants and to *never handle tobacco while working with tomatoes,* as tobacco spreads the virus. It is a good idea to wash your hands in a milk solution before touching tomato plants where the virus is a clear danger—milk deactivates the virus. Another control is to spray each twenty square yards of plants with a solution made of one gallon of milk and one gallon of water. Moto-Red is a resistant strain.

Verticillium Wilt. A common soil-borne fungus disease that attacks the leaves and vascular systems of the tomato plant, weakening it and affecting bearing. The chief symptom is yellowish splotches on the lower leaves in the center of the plant. Soon chocolate-brown spots develop in the middle of the yellow area and the plants drop some lower leaves. Midday wilting and evening recovery is common. Careful inspection of a slit stem near the soil line will show tan streaks just under the skin. This dead tissue prevents proper transport of food and minerals to and from the root system. Controls are to plant in ground free of the verticillium fungus, or to plant one of many verticillium-resistant varieties. Keep the garden clean and don't dig potatoes or tomatoes, or their foliage into the garden, as they can harbor the disease. Chemical fungicides do little good here because the verticillium fungus lives so long in the soil.

Walnut Wilt. See Chapter Eight for this wilt condition caused by planting tomatoes close by black walnut trees.

Weed Killer Injury. Tiny amounts of common weed killers like 2,4-D and 2,4-5-T can be injurious to tomato plants, causing leaves to twist and fruit to become cracked, catfaced, or cone-shaped. Take great care when applying these weed killers near tomatoes; in fact, try not to do so at all. Stricken plants can be nursed along with water, then fed, but this doesn't always work.

Just the most common tomato diseases have been listed here. If you want to pursue the subject, and study the causes and symptoms of diseases more deeply, there are a number of booklets on the subject. *Market Diseases of Tomatoes, Peppers and Eggplants* (45¢) and *Controlling Tomato Diseases* (free) are available from the Office of Information, U. S. Department of Agriculture, Washington, D.C. 20250.

Chapter Fourteen

THE "OUTTA SPACE" TOMATO PATCH: TWENTY-FIVE SPACE-SAVING WAYS TO GROW TOMATOES

Last summer a Georgia gardener grew 230 tomato plants in exactly *one square yard* of garden space. He accomplished this spectacular feat by building a twenty-seven-foot tower spiked with planting holes, filling it with rich compost, and fitting it with a long perforated pipe down the center that he could water through. Picking the fruit did prove to be something of a problem, but this was more than compensated by an "outta space tomato patch" equal to about eight hundred feet of conventional garden that yielded hundreds of tomatoes.

If you think that tomato patch way up in Athens, Georgia, is way out, then look out, for a tomato patch may be coming at you down the highway. One California man actually grows tomatoes atop his Volkswagen, which he drives all over town. The car roof was bashed in with a sledgehammer so that it formed a receptacle for soil and four tomato plants were set in it. The driver estimates that a ratio of six hours street use for every one hour of care promotes the best growth, and he never plants on the front hood, as that would obstruct visibility!

Another enterprising tomato gardener, a little less unorthodox, converted the strip of "useless" clay (eighteen inches wide by forty-two feet long) between his driveway and his neighbor's into a productive "tomato strip." By mixing in leaves, grass clippings, and manure with the clay he was able to harvest bushels of tomatoes from his "vertical garden."

There you have them—horizontal, vertical, and mobile tomato patches that innovative green thumbs have created across the country. And by no means do tomato space-saving innovations end with these ingenious growers. Other ploys include land rental, the utilization of front lawns, the creation of more backyard space, and container-planting in apartment house window sills or in "sky farms" fifty stories above street level. Let's take a closer look at some of them.

Above left. Fruit-laden Presto, grown and ready to be eaten on the patio.

Above. Tumbling Tom does especially well in hanging baskets, but is also suitable for patio containers and window boxes.

Left. This prolific Patio Tomato makes every patio a tomato patch.

RENT-A-PATCH: LEASING LAND FOR TOMATO GROWING

In England gardeners have for hundreds of years worked annual "allotments" —small parcels (about three hundred square yards each) of vacant public land that they can rent for $3.00 or so if they put it to productive use. This ancient British tradition dates back to the enclosure acts of the seventeenth and eighteenth centuries, which deprived villagers of common grazing rights. The poor were compensated for their loss with the "allotments" and today there are over half a million such plots in Britain. It's true that the British are a fanatic race of gardeners; not long ago, a garden columnist for the *Times* whose cucumbers had been insulted by a *Daily Express* writer challenged his colleague to "a duel with marrows at twenty paces." But there's no reason why their allotment system wouldn't work in America. In fact, Americans are already renting public and private lands for tomato patches and gardens of all kinds. Here are a few ways people have rented garden space across the country:

- Several Indiana cities offer garden plots to the public—$25 rents a 20-by-25-foot plot.
- The Married Students Council at Indiana University rents small plots to couples for $3.00 a year, the land leased from a nearby farmer.
- An Illinois nurseryman rents tomato patches on his premises.
- An industrial concern in Columbus, Ohio, bought land nearby that it rents to employees who want to start small gardens.
- In Windsor, Connecticut, the city fathers provide space to green thumbs in a five-acre Fight Inflation Garden behind the town's community center.
- Rubble-filled lots in Manhattan are rented from the city for as little as $1.00 a year, cleared and planted with vegetable gardens. (The fee even includes use of water from fire hydrants.) The non-profit Park Council of New York (80 Central Park West, New York, New York) will give guidance and assistance to anyone interested.
- In Appleton, Wisconsin, a local farmer rents a thousand square feet of his pasture (already plowed and fertilized) for only $10.
- Huntington, New York, residents can rent 20-by-40-feet plots for only $5.00 in a strictly organic community garden. The rental fee includes use of a tiller and generous supplies of manure, mulch, and water.
- In many communities throughout America gardeners rent vacant lots from private owners whose land is lying idle.

If you decide to rent land for a tomato patch from either the city or a private party, there is a definite procedure to follow. This varies from place to place, and may be difficult, but persist in working through the maze of government red tape to obtain permission for land use. In New York and other cities, it works this way:

1) Note all the street names surrounding the lot you choose, and the numbers of adjoining houses.

2) Go to the borough Real Estate Registry (listed in the phone book) and consult the map that gives block, lot, and index or parcel numbers. From the Deed Index find the name of the deedholder, or contact the owner through the lawyer listed on the deed.

3) If the lot is city-owned, a letter should be written to the Commissioner of the Department of Real Estate. Describe the project planned; give block, lot, and index or parcel numbers.

4) The lease for the lot should be in writing.

5) Before proceeding with any gardening, it is advisable to obtain "third person" liability insurance. With city land this is almost always required.

"We live in a grass country; we live in a grass world," a New Rochelle, New York, gardener lamented not long ago, pointing an accusing finger at a lush lawn across the street. "Do you ever see anyone walking on lawns like that or sitting on them? Of course not—all that green, inedible stuff is totally useless!"

PLANTING TOMATOES ON THE FRONT LAWN

Many Americans are beginning to think like the New Rochellian and have, like him, planted their front lawns in vegetables. One Texas man actually planted *wheat* on his front lawn—enough to make one hundred loaves of bread—and persevered even though his neighbors complained to City Hall! There's no law against using your front lawn to plant any crop in most communities, and if there is, you might be able to have it changed. The plush front lawn that yields harvests of bills, worry, and backaches is strictly a modern invention and one that often makes little sense.

Gardens planted on the front lawn not only save money and reward one with healthy, tasty food, but can be attractive as well. Use a little imagination and a big garden or a small tomato patch will blend in nicely with other plantings. As for the work involved, it can be minimal. If you don't want to turn over the front lawn, take a short cut and mow the grass close, spreading about three inches of decayed leaves over the portion of the lawn to be used. After this compost has settled down, simply make planting holes for the tomatoes and set them in, two feet apart each way. Try to use attractive mulches such as stones or gravel. Plant marigolds to combat the meadow nematode and care for the plants just as you would in the backyard. The turf grass, smothered, will soon become instant compost and yield bushels of delicious tomatoes.

CREATING MORE
BACKYARD SPACE
FOR TOMATOES

Even a postage-stamp backyard will grow some tomatoes. Try any of the unique ring-staking systems described in Chapter Nine—the Chinese ring system, for example, will yield lots of fruit in a relatively small space. Tomato plants can also be spaced closer together than is usually recommended where space is a problem; only experience will tell just how close you can plant in your backyard. And tomatoes will yield, though sparsely, in partial shade. (See Chapter Three.) Other space-savers include growing plants on a fence or training them to grow up a wall of the house or garage. Then there are the "tomato towers" mentioned previously, and tomato pyramids made similar to the strawberry pyramids common in many gardens.

A space-saver often overlooked is growing tomatoes in the flower garden. The plants are easily combined with annuals there and make an attractive display in addition to providing food. Novelties like striped, white, and green tomatoes, or distinctive plants like the bronze-leaved Abraham Lincoln described in Chapter Two, are particularly effective. Be sure to place tomatoes and flowers of the same height together. For the garden background, combine staked hybrid tomatoes with tall annual flowers; for the middle of the border, smaller indeterminate-tomato types are best; and the smallest cocktail tomatoes should be used for edging plants.

CONTAINER
PLANTING:
GROWING
TOMATOES
WITHOUT
A GARDEN

Over 35 million Americans live in apartments, and many more have no room whatsoever in the yard for tomato growing. But there's no reason why apartment dwellers and others without planting space can't harvest fresh, garden-ripe tomatoes. Anyone with a windowsill, balcony, patio, fire escape, roof, porch, or doorstep has space enough for a tomato minigarden. In fact, of all vegetables, tomatoes probably offer the largest edible reward for the time and effort spent in container gardening. Start a container garden by picking out the sunniest accessible spot available—remember that tomatoes *ideally* need at least six hours of sun a day—and then follow these steps:

1) Container selection. The large containers in which tomatoes can be grown to maturity are only limited by the imagination. Following are just a few of the containers that have been used for tomato planting: bushel baskets, plastic laundry baskets, metal pails, clay and plastic pots, trash containers, plastic mop pails,

Climbing tomatoes and other vigorous varieties can easily be grown on a trellis.

The tree tomato, a close garden-tomato relation, makes an interesting, productive patio plant or houseplant.

large plastic sacks, dishpans, wooden barrels, window boxes, large earthenware urns, wooden butter tubs, wastepaper baskets, wine casks, wheelbarrows, strawberry barrels, hanging baskets, large plastic milk containers, vegetable crates, and redwood boxes.

Many gardeners construct their own tomato containers from redwood, cedar, or cypress (which don't rot easily) or even from ¾-inch plywood. Others build permanent containers of brick, cinder block, concrete, or tile. Containers holding tomatoes can be very attractive on a patio or terrace, or simply plain and utilitarian. Even a large plastic bag filled with planting mix can be used—it is just tied after being filled, laid down on one side, and two-inch holes are cut in it to receive two or three plants.

What is most important from a growing standpoint is that any tomato con-

A plant raised to maturity in a plain clay pot.

tainer should be big enough for the plant it contains. Containers for small variety tomatoes should be able to hold at least one gallon of soil. (This would be equivalent to an eight-inch clay pot.) For large variety tomatoes use at least two-gallon capacity sizes (about the equivalent of a ten-inch clay pot). For the rampant growers, the bigger the container, the better.

All tomato containers must have large drainage holes in the bottom and should be lined with broken pieces of shard, charcoal, small stones, or a piece of aluminum wire screening so that soil doesn't leak out through the drainage holes. Never set a container on the terrace floor, which will bake plant roots on hot days; instead, set the planter on bricks or boards so that air can circulate beneath it.

When building a wood planter for tomatoes, use galvanized nails or brass screws. Fasten two strips of wood on the bottom (one on each end) for the planter to rest on—this will help with drainage and air circulation. To preserve cheap, fast-rotting woods, do *not* coat the inside with creosote, which can kill plants. Use either Cuprinol, asphalt compound, or a quick-drying outdoor house paint. Lining planters with copper is an alternative, but is expensive.

A final cautionary note: Be sure that any window-box planter or hanging container is securely fastened. A window box with soil, for example, can easily weigh five hundred pounds and take someone's head off if it falls a few floors to the street below. Always check with the building superintendent before installing any planter, and have a competent carpenter or handyman check out your plans if you're installing a planter off the ground at home.

2) Soil for Containers. There are many commercial potting mixes available for filling large containers, or a number of do-it-yourself formulations can be used. Three of the latter follow:

▪ Compost Mix: Mix together three gallons of good soil, three gallons of compost, two gallons of builder's sand, a half pint of cottonseed meal.

▪ Soil Mix: Mix four gallons of earth, one and a half gallons of sphagnum moss, one and a half gallons of builder's sand, one gallon of dried cow manure, one pint of cotton seed meal.

▪ USDA Soilless Fertilized Mix: To one bushel each of horticultural grade vermiculite and shredded peat moss, add fourteen ounces of ground dolomitic limestone, four ounces of 20 per cent superphosphate, and eight ounces of 5-10-5 fertilizer. This material should be mixed thoroughly. If the material is very dry, add a little water to reduce the dust during mixing.

The USDA mix is the best of the lot and easiest to work with. Its lightness makes it by far the best soil for hanging baskets or window boxes. But whichever

*Toy Boy, a new
introduction especially
adapted for hanging
baskets.*

formulation you use, be certain to fill the container with moistened mix up to one inch of the rim.

3) Varieties. Over thirty midget varieties suitable for container planting are listed in Chapter Two and many more small plants can be found in Appendixes I and II. Small Fry, for one, is excellent staked in a small container, as it grows about three feet tall. Tiny Tim, which grows only eight to fifteen inches tall, does well falling naturally from a hanging basket, as does Toy Boy. All larger tomatoes can be grown in containers, too. Just be certain that their containers are big enough—at least 2-by-2 foot containers are needed for the giant varieties.

Several tomato varieties will grow well in hanging baskets provided the container will hold at least two quarts of soil. Place drainage material in bottom, then add appropriate soil mixture. Potted plants at right are the correct size for transplanting.

Press the young tomato plant firmly into the surrounding soil so that the roots take hold.

Always water the container thoroughly. A solution containing one fifth the fertilizer recommended for houseplants is best for these tiny tomato plants.

A small hanging basket, such as this one, can yield twelve to fifteen fruits.

Pixie Hybrid is an excellent small-fruited plant to grow indoors in a sunny window.

4) Planting and Staking. Plant tomatoes, either raised from seed or bought from a nursery, in the same manner as you would set transplants into the garden. (See Chapter Eight.) Stake all plants except small varieties that will grow in hanging baskets. Almost all the staking methods given in Chapter Nine can be adapted to container planting; i.e., a "tomato tree" can be made in a basket by using a cylinder of concrete reinforcing wire . . . the plants can be trained on trellises or tepees . . . plants can be espaliered to a stake made in an espalier form, etc. Staking will provide more room for plants to grow and they'll be easier to care for and be far more attractive.

5) Watering. Tomato plants in containers need about one inch of water a week and must be watered whenever the soil becomes dry down to a depth of an eighth inch. This means watering perhaps three times a week during hot, dry weather. Don't overwater, however, or you'll slowly drown your plants. When using a sprinkler, never water so late in the evening that the leaves of plants stay wet at night, which encourages plant diseases.

If you want to be absolutely sure that water reaches the bottom of the container, insert a pipe or piece of drainpipe punctured with holes into the pot. The pipe should reach from the soil line to the bottom of the container. Fill it with builder's sand and pour water into it so that water will always reach the roots.

Where obtaining water is a great problem, try to use drought-resistant varieties like Calmart and give as much water to the plants as you can. Only by giving this approach a try will you determine whether you can grow tomatoes in a particularly dry location.

When watering container plants, follow the same general rules for watering and mulching given in Chapter Eleven. There is, however, a new method of watering developed by the USDA *specifically for plants in hanging baskets.* The USDA recommends that these plants be watered as soon as they are set in the baskets with a general houseplant solution *at one fifth its recommended rate.* Plants should not be watered again until they begin to turn pale green and wilt. From then on water only every three to seven days—when the plants are wilting. This watering method will toughen the plants and make them better able to adapt to basket life.

6) Fertilizing. The fertilizer in the soil mix will support plant growth for about six weeks. After that either use the USDA watering method outlined above for hanging baskets, or feed plants once a week with a solution of one tablespoon of water-soluble fertilizer per gallon of water. Be careful not to use too much nitrogenous fertilizer. (See Chapter Ten.) If you want to use dry fertilizer, feed each plant one level teaspoon of 5-10-5 fertilizer per square foot of soil every three weeks.

Tumbling Tom does well indoors in a hanging basket if given enough sun.

Harbinger tomato, an English variety, growing in ten-inch pots in a greenhouse.

7) Sunlight. Remember that three hours of sunlight on one part of a balcony in the morning and three hours of sunlight at the opposite end in the afternoon do not add up to the recommended minimum of six hours of sunlight for each plant, not if plants aren't moved daily to follow the sun. Always move plants when possible to get maximum sunlight. If a plant is too heavy to be moved and has its back against a wall or other object blocking the sunlight, turn the container at least once a week so that the plant develops symmetrically and is well balanced.

8) Pollination. If flowers appear on terrace plants but they don't set fruit, there may not be enough wind to effect pollination. Gently shake the plants once a day to remedy this.

9) General Care. Container-grown tomato plants are subject to the same diseases, insects, and disorders as plants grown in the garden, and should be cared for in the same way. Weed the plants regularly, watch them for signs of diseases, feed and water them, and they'll soon be bearing more tomatoes than you thought possible.

Chapter Fifteen

TOMATOES UNDER THE CHRISTMAS TREE EXTENDING THE TOMATO SEASON: OUTDOORS, INDOORS, AND IN THE GREENHOUSE

You can easily stretch the outdoor tomato season many weeks beyond the first fall frost date in your locality—even up to Christmas—by using the same techniques described in Chapter Seven for early tomatoes. Just how long the plants will bear depends on the severity of the weather, but the use of heavy plastic and other protective devices will certainly insure tomato production long after most gardeners nearby have pulled out their plants.

Old-timers say that the full moon of an autumn month invariably brings frost and that then the weather warms again until the next full moon. Whether you rely on such ancient adages (which are often correct), or on modern weather forecasting, always cover late-season tomato plants at night when a frost is expected, using protective plastic sheets, baskets, or other materials. Light nighttime frosts won't kill the plants but *will* harm tomatoes on the vines; sunrays are magnified by the frost the following morning and burn them, which leads to their eventual decay.

Many gardeners extend the outdoor tomato harvest all the way up to Christmas by using a cold frame as noted in Chapter Seven. They simply layer a few branches on firm, healthy plants in mid-September (see Chapter Eleven Propagating Tomatoes) and when these protected branches have rooted into vigorous new plants by November, they are transplanted to a cold frame filled with small sections of cornstalks. The cornstalks, cut into six-inch pieces, are mixed with cottonseed meal, watered heavily, and tramped down about a month before plants are set in them. At planting time they have started to break down and generate heat, and are covered with a one-inch layer of straw plus four inches of compost. Thus when the tomatoes are set in the cold frame (or hot frame, as it would be more correctly called now), they will receive about six weeks of gentle bottom heat to keep them growing and producing until Christmas.

It might even be possible to keep plants alive until the following spring by using the above method—if they were packed in straw and the cold frame was carefully covered at night—but the plants wouldn't produce tomatoes after the first of the new year. All-year-round tomatoes in most areas must be grown indoors for the winter. This can be accomplished in the following ways.

GROWING INDOOR TOMATOES IN A SUNNY WINDOW

Bushel baskets of tomatoes won't be harvested from plants grown in a sunny window, but the fruits that do grow will be all the more delicious because of their scarcity. And it's safe to say that you'll be one of the few in your circle to have "edible houseplants" if you observe the rules below.

Choose Indoor Varieties. It is important *not* to choose a vigorous variety for indoor growing. Much more success can be achieved with small plants. Pixie Hybrid, Atom, Patio, Toy Boy, Tiny Tim, and Small Fry are all excellent choices; plants of Atom grown in a sunny window have produced a winter crop of six hundred tomatoes each. Good, too, are all the small cocktail tomato varieties noted in Chapter Two. If you do choose to grow larger plants, don't try anything larger than medium-sized tomatoes like Marglobe, Rutgers, and Fireball. Varieties bred for greenhouse growing are also fine for a sunny window.

Starting indoor plants. Either start plants from seed indoors early in September, using the methods outlined in Chapter Six, or bring plants in from the garden. Never select huge fruit-laden garden plants and attempt to transplant them to pots. Such plants have enormous root systems and will be harmed beyond repair when dug up. Instead, either layer new plants from old ones, take slips from old plants (see Chapter Eleven), or select small volunteer plants that may have sprouted from seed dropped in the garden. Another option is to sink several small potted plants into the garden soil well after plants are set out in the spring and take them inside, pots and all, when cold weather comes. All movable patio plants can also be taken inside. Still another method is to seed tomatoes directly into the garden in early August and pot these plants for indoor use before frost comes.

Containers. Most of the containers mentioned in the preceding chapter can be used for potting indoor tomato plants. The main difference is that they must be set in a tray, pan, or dish to catch drainage water. Rest the tray on a mat or a magazine so that it won't mar the finish it rests on. The pot in the tray should be raised slightly from the bottom by resting it on small stones, marble chips, or strips of wood, so that the pot doesn't sit in water.

Planting Mix. Of the container mixes mentioned in the last chapter, the USDA vermiculite mix is best for indoor tomatoes. But prepared soil mixes like African Violet Soil are good, too.

Light. This is the most important factor in indoor tomato growing. A sunny window facing south is almost essential for any real success. An east window is the second choice and a west window third best, though plants will hardly bear in these two locations. A north window is virtually useless.

On dark days you can give your plants an extra dose of light by putting them about a foot away from ordinary lamplight. Incandescent lights don't have sunlight's range of light waves but will help some.

Heat. Maintain a temperature of from 65° F.–75° F. during the day and 60° F.–65° F. at night. Space plants about a foot from the window and pull the shade or drapes at night to give them more warmth. Do not put the plants close to a radiator or heater. Too much direct heat will harm them more than cold.

Ventilation. Be sure that there is good air movement in the room, even if you only open a window or door a few minutes every day.

Watering and Feeding. Water and fertilize plants as described for outdoor container plants in the preceding chapter. If you are away for any length of time, have someone water the plants, or buy a glass-wool wick system for automatic watering.

Pruning. When small varieties are used there will be little need to prune. If plants do grow too tall, you'll have to prune every week, pinching back new growth so that the tops don't grow out of balance with the relatively small root systems in the containers.

Pollination. Tomatoes are self-pollinated and indoors have to be tapped firmly or even vibrated with an electric toothbrush when new blossoms appear so that the pollen will scatter. You can also spray the blossoms with a fruit-setting hormone. Outside, winds and insects take care of pollination.

As a final tip remember to rotate indoor tomato plants in the window at least once a week. The plants will grow more symmetrically this way. A New Zealand researcher has found that rotating plants counterclockwise stimulates their growth, while clockwise rotation inhibits it.

GROWING TOMATOES INDOORS UNDER ARTIFICIAL LIGHT

Raising winter tomatoes under fluorescent lights like Gro-Light or Gro-Lux is the only answer for those who don't have a sunny enough window. And plants grown under artificial light will produce far more fruits than those growing in any window, no matter how sunny. The lights can be set up anywhere in the house— in a vacant closet, the basement, the attic, on living-room wall shelves. . . . Or

"Moneymaker" tomatoes growing directly in the soil in a greenhouse.

Greenhouse plants growing on wooden stakes.

you can use carts with built-in lights that can be moved from place to place—
even into the hall of an apartment building.

Except for the light source, treat plants grown under fluorescents the same as
plants growing in a window. The light should be at least a pair of 40-watt fluorescents four feet long for each eight square feet of space. Keep the plants about six
inches from the lights. Be generous with the light, too, giving plants sixteen to eighteen hours of it daily. Automatic timers are available that will enable
you to do this more consistently when at home or away.

It's also a good idea to buy a wide reflector that attaches to the grow-lights and
focuses more light onto the plants. White or aluminum painted backgrounds
reflect light—well, too—as do small mirrors placed among the plants. More information on artificial lighting systems can be found in Chapter Six.

GROWING
TOMATOES IN
THE GREENHOUSE

Anyone fortunate enough to have a greenhouse can grow tomatoes in it all
year round. You'll be most successful if you use tomato varieties bred specifically
for greenhouse culture, especially the small-fruited types among these. Below is a
list of particularly good greenhouse tomatoes, which can be supplemented by
many other varieties given in Appendixes I and II:

All small-fruited "cherry" types	Michigan-Ohio
Bay State	New Yorker
Burpeeana Early Hybrid	Ohio-Indiana
Caro-Red	Parks Extra Early Hybrid
Doublerich	Rutgers*
Fireball*	San Marzano*
Floralou	Toy Boy
German Dwarf Bush Imp	Tuckcross
Heinz 1350	Vantage
Homestead	Veegan
Manapal	Vine Queen
Manalucie	Vendor
Marglobe*	

Care for greenhouse tomatoes is more difficult than for plants grown outdoors.
Good sanitation and ventilation are especially important—many diseases will be
a persistent problem if they are not provided. Though experience is the best teacher

* Especially suitable for taking slips or layering in the outdoor garden in August to make new
plants to grow indoors over the winter. San Marzano, however, must be pollinated by hand
indoors.

These tomato plants growing in a commercial greenhouse are supported by string "stakes" that they are wrapped around as they grow. This technique can be used equally well in your own greenhouse.

Heinz tomatoes can be grown successfully in a greenhouse as well as in the garden.

in greenhouse, following pointers will help you grow better tomatoes under glass:

- Always pasteurize greenhouse soil. (See Chapter Five.)
- Select greenhouse varieties or disease-resistant types for greenhouse growing.
- Use only treated seed, or treat seed saved from the garden yourself. (See Chapter Sixteen.)
- Rotate greenhouse tomatoes every year, never planting them in soil where tomatoes, potatoes, peppers, or eggplants grew.
- Most growers of greenhouse tomatoes produce a spring and fall crop. The spring crop is easier to grow and yields are considerably higher. This is due to the improved light and other conditions in the spring and early summer compared to the fall and early winter. The spring crop is seeded in late December or early January and the fall crop is started in late June and early July.
- Start all seed as described in Chapter Six.
- On an overcast day, transplant seedlings to any container, so long as it is large enough for the variety. (See Chapter Fourteen, Container Planting.)
- The best temperatures for growing tomatoes in the greenhouse are 70° F.–75° F. in the daytime and 60° F.–70° F. at night. Don't let daytime temperatures exceed 85° F. or nighttime temperatures fall below 60° F.
- Make certain that the greenhouse is properly ventilated. Open the vents on hot days and follow all other pertinent directions for greenhouse operation as noted in your owner's manual.
- Fertilize tomato plants once every two or three weeks with a balanced liquid fertilizer.
- Water plants whenever the soil is dry to the depth of about an inch. Don't splash plants with dirt and never water the foliage on greenhouse plants, especially late in the day when they won't have a chance to dry off by night.
- To conserve room, prune plants to a single or double stem as described in Chapter Nine. Stake vigorous varieties just as you would outside. Another technique is to stretch wire between two stakes seven to eight feet high over a row of plants. Then drop string from the wire to each plant so that it is loose enough to allow for plant growth. As the plant grows carefully train it upward by twisting it around the string.
- Greenhouse tomatoes can't be insect or wind pollinated, so they must be pollinated by artificial means. In late morning, on a bright day when the plants are dry, shake those plants with blossoms on them or vibrate them with an electric toothbrush.
- Practice the same good sanitary habits in the greenhouse that you would in the garden. Especially don't smoke when tending plants, as this can spread the tobacco mosaic virus.

The Greenhouse Handbook ($1.25), available from the Brooklyn Botanic Garden, 1000 Washington Avenue, Brooklyn, New York 11225, is a good guide for growing plants under glass, but there are many books on the subject. *Under Glass* magazine, published by Lord and Burnham Company, Irvington, New York, is an excellent periodical to subscribe to. If you're interested in growing large crops of greenhouse tomatoes for the local market, *Greenhouse Tomatoes* ($4.75) is a complete guide and can be ordered from the Michigan State Press, East Lansing, Michigan 48823.

A wide selection of greenhouses are available today at prices ranging from about three hundred to thousands of dollars. If you want to build your own a lot cheaper, the best free bulletin on the subject, complete with designs, equipment, costs, and sources of plans, can be had by writing for "Home Greenhouses, Circular 879," from the Agricultural Information Office, 112 Mumford Hall, University of Illinois, Urbana, Illinois 61801.

Chapter Sixteen

REAPING THE REWARDS:
HARVESTING TOMATOES AND SAVING SEEDS

Country lore instructs that the best time to harvest tomatoes is during the decreasing light in the third and fourth quarters of the moon, which will insure easier gathering and better preservation of the crop. Harvest in the *increasing* light phases of the moon, we're told, and fruits will bruise and rot easily. This may or may not be true, but it is a lot simpler to depend on your eye than moon phases. Tomatoes taste best and are most nutritious when eaten dead-ripe straight from the vine, their flesh warm in the hand. Peak of ripeness usually comes about six days after the first color shows. Tomatoes will stay in this dead-ripe condition on the vines only two to three days before beginning to decay, so you must keep a sharp lookout in the garden when fruits are ripening. Dead-ripe fruits are always strong in color (bright red, orange, etc.), full and shiny. They can be stored ripe in the refrigerator, of course, but begin to lose in flavor and nutritive value the moment they are picked.

You can, of course, help tomatoes to ripen on the vine by using one of the many protection measures described in Chapter Seven. Or you might want to use an old trick described by a nurseryman in the 1890s: "At the approach of frost the plants will be loaded with full-sized fruits just beginning to put on the first whitish tinge—the first indication of ripening. In a warm situation, with northern protection, dig a 30″-deep trench wide and long enough to contain the plants, which should be cut quite close to the ground. Spread out the plants with their green fruit in the trench until about two feet thick, and over them place a covering of straw six inches in depth, which should be held in place by the use of some light bush. The warmth from the earth will ripen the larger fruits perfectly."

But sometimes it is impossible to pick tomatoes ripe from the vine. You may be away for an extended period, or an early killing frost might be predicted before there is a chance to protect the plants. Don't despair if this happens. Partially unripe or green tomatoes can be ripened in several ways. Here are a number of them:

1) Ripen large green tomatoes indoors at temperatures between 60° F. and 72° F. Temperatures below 60° F. delay ripening, while temperatures above 72° F. are likely to cause undesirable color and decay. Light is not necessary, although it will increase the color of tomatoes somewhat. In any case, don't put tomatoes in direct sunlight indoors—the added heat often deteriorates their quality. A north window is best if green tomatoes are ripened in a window. Make sure that the fruits aren't cracked or blemished, and that they don't touch, so that any possible decay can't spread from one fruit to another.

2) Store green tomatoes wrapped in newspaper or packed in individual containers. If the fruits are kept at a temperature of 55° F. using this method, ripening will be slower and they'll last longer. Green tomatoes can be disinfected with a solution of one teaspoon of household bleach to a quart of water before being individually packed. Dry them thoroughly with paper towels before storing.

3) Store tomatoes separately in plastic bags, making small holes for ventilation.

4) Leave a few inches of stem on each tomato when taking the fruits from the vine and store them on trays or racks in a single layer.

5) Pull up an entire tomato vine with green fruits attached and hang it stem up anyplace where the temperature is always between 55° F. and 72° F.—a basement, attic, garage, spare room, etc. An easy way to do this is to string a rope from wall to wall and hang the plant on it. The fruits will continue ripening long after the plant has wilted.

MORE TIPS ON TOMATO STORAGE

▪ Remember that blotched fruits should not be stored but used immediately, for they quickly rot.

▪ Very small, dark-green tomatoes will never ripen and must be either pickled and preserved, or made into piccalilli, chutney, or any of the green tomato recipes given in Chapter Eighteen.

▪ Other green tomatoes stored at 65° F.–72° F. will ripen in about two weeks; green tomatoes stored at 55° F. take three to four weeks to ripen.

▪ A new variety called Thessaloniki is a tomato that keeps for a long time even if picked when fully ripe.

▪ All tomatoes stored for ripening taste better if they are scalded with hot water and their skin is peeled off.

Usually it doesn't pay to save tomato seeds from fruit that ripens in the garden or from a particularly delicious tomato you happen to eat. Many gardeners do this, but there are a number of problems you'll encounter even if all the rules given here are followed. The most important is that seed from *hybrid* plants rarely produces the same plants that it came from.

Hybrid tomato plants are one of the great developments of the vegetable world —extremely hardy, prolific vines that produce high-quality fruit all season. Any hybrid plant is the offspring of a cross-fertilization between two parents different in one or more genes. But tomato plants have both male and female parts in the same flower, pollinating themselves. Therefore, in crossing, pollen must be taken from "male" Tomato X and placed on the pistil of "female" Tomato Y. All flowers of the mother plant Y must then be emasculated (the thin pollen sacs removed by hand) before being pollinated with the male pollen from Tomato X. The seed obtained from this cross produces hybrid F1 plants that often grow better, yield more, have better disease resistance, etc. What the plant breeders have done is to "breed in" good qualities of each plant, "breeding out" poor or weak qualities.

Seed from the first generation of an F1 hybrid (which produces F2 hybrid plants) will result in plants similar but not identical to the F1 hybrid. These plants will generally not be as good as the original, and if you save seeds from them (the F2 plants), the F3-generation plants that result will be even less like the original F1. F2s and F3s are usually "lost generations." The point is that because hybrid seed does not come true from seed, it is obviously a waste of time saving it. No seed really comes completely true to form if the plants are closely examined. Any plant from a seed is a unique individual different than its parents, and identical individuals are only produced from rooted cuttings of a plant or when a cutting is grafted onto another stock. But hybrid seed produces progeny with more than slight variations. Nevertheless, many people don't care what kind of tomatoes they get and plant seed from any tomato at all. In fact, some green thumbs do nothing but plant seeds from a particularly good winter tomato they've eaten and transplant the seedlings into the garden when they're big enough, not even bothering to sterilize the seed. But St. Fiacre looks over these people, and when the patron saint of gardeners is busy elsewhere their luck can run out, resulting in a tomato patch wiped out by a seed-borne disease.

If you do want to save tomato seed from the garden, follow the rules of the experts to the letter and save a lot of trouble. To begin with, eschew all hybrids and stick to old standard varieties, whose seed will produce nearly identical offspring. Actually, I'd even go a step farther and say that tomato seed should be saved only from old standard family varieties that can't be bought from seedsmen anymore, but which you want to perpetuate. Tomato seed is too inexpensive to bother with the time-consuming process of saving it for any other reason. (See also Appendix III.)

PREPARING AND
TREATING
TOMATO SEED

When saving old family varieties, or any tomato seed from the garden, select the best tomato on the healthiest vine you have. Pick the fruit when it's dead ripe (but not overripe) and scrape the seeds out with a knife. Next soak the seeds for three days in a pot of water at room temperature to allow them to "ferment." Stir the mixture several times a day and finally pour off whatever pulp and dead seeds are floating on top. (The seeds with life in them will have settled on the bottom.) Then rub the seeds until the remainder of the pulp comes off—this is important, for seeds with flesh on them are apt to rot. After the flesh has been removed, wash the seeds in cold water and dry them on sheets of paper in a dry, warm, well-ventilated room, taking care to turn them over periodically to prevent formation of mold on their undersides.

Once the tomato seed is dry, treat it to help prevent losses due to soil micro-organisms, insects, or seed-borne diseases. This procedure can be combined with cleansing the seed but is given separately here because it is a good idea to process commercial seed, too, if it hasn't been treated. (Examine all seed packets to see if the seed has been treated.) If such is the case, simply immerse the seeds in a pot of water at 122° F., stirring with a paddle and letting the seed soak for at least twenty-five minutes before drying it. Do the same with seed saved from the garden. This heat treatment will kill bacteria and fungi responsible for many diseases but won't kill the seed.

As an added disease precaution, you can lightly coat tomato seeds with thiram, a chemical dust that provides good protection against seed-rotting fungi in the soil. Do this after the seed is dry by dipping the top of a knife in the dust, flicking it into the seed packet, and shaking it. Only a very light coating is necessary. Store the thiram and treated seed in a safe place beyond the reach of children and pets.

STORING
TOMATO SEED

Ideally, tomato seed from the garden, or from partially used seed packets, should be stored where the temperature is 40° F.–50° F. and there is a low relative humidity of 45 per cent or less. However, tomato seed will last up to five years in a cool dry place out of direct sunlight. If the seed is in partially used commercial seed packets, reseal the packets with tape or staple them closed after folding over the top edge several times. Seed gathered from the garden can be kept in sealed envelopes, coffee cans with plastic airtight lids, empty baby food jars, or in homemade paper packets taped or stapled closed at the top. Some gardeners store their seed in the refrigerator in a fruit jar, along with a little bag of desiccant to absorb moisture. Where vermin are a problem, put any paper packets in a metal box as an added precaution. Be sure that the seed containers are adequately marked with the variety name and the date of storage.

If the date on a tomato seed packet is for the current planting season, there's no need to test it for germination, but if the packet is an old one or you use garden seed saved from last year, try the "rag doll" germination test before planting. Just place ten or so seeds a half inch apart on a moist cloth about a foot long and a foot wide. Roll up the cloth, covering the seeds, tie the ends of this "rag doll" with string or rubber bands, and place it in a plastic bag (pinholed for ventilation) to prevent it from drying out. After five to seven days pass, untie the "doll" and count the number of sprouted seeds. Divide these by the number of seeds in the rag doll, then multiply by one hundred, and you will have the percentage of germination: i.e., eight seeds sprouted divided by ten seeds in the doll equals .80×100, or 80 per cent germination.

The same formula applies if seed is tested in a damp paper towel kept moist between two plates, pasteurized soil, or any other sterile planting medium. If more than 70 per cent of the seeds germinate, it is safe to assume that the seeds are quite viable or good to use. If less than 50 per cent sprout, you may want to try sowing them thickly when planting, but don't bother saving the seeds for the following year.

Chapter Seventeen

PUTTING THEM UP:
SALTING, DRYING, FREEZING, AND
CANNING TOMATOES

The good qualities of tomatoes are less destroyed by canning than those of any other fruit or vegetable. But canning isn't the only way to preserve a tomato surplus for a taste treat during the long winter. Tomatoes can also be salted away ripe, dried, and frozen. Whichever method is chosen, try to use only the best of the crop—not diseased, overmature, or immature fruits—and to process tomatoes as soon as possible after harvesting, preferably the same morning or afternoon.

SALTING THEM

This is a little-used technique, but you might experiment with it using a few tomatoes. Sprinkle a layer of coarse salt in the bottom of a large wooden barrel. Then add a layer of firm, ripe unblemished tomatoes, leaving at least one inch between each fruit so that they don't touch each other. Cover the tomatoes with another layer of salt and continue the alternate layering until the barrel is full or you run out of tomatoes. The salt does not affect the taste of the fruit, which will remain fresh and firm for a long while.

KEEPING
TOMATOES UNDER
REFRIGERATION

Keep ripe tomatoes in a refrigerator at a temperature of 40° F., taking care that they do not touch. They will last four to six weeks this way. Check periodically and remove any spoiled fruits.

DRYING TOMATOES

There is some loss of vitamins A and C in dried tomatoes, but drying is one of the easiest preservation methods. Frequently, gardeners cut tomatoes into quarter-inch slices and lay them out in the sun to dry. But there are more foolproof ways to dry "tomato chips" for use in soups, stews, and even as a snack. Another easy way is to cut firm, ripe, perfect tomatoes into quarter sections, squeeze out as much of the juice as you can, and remove as many of the seeds as possible. Then sprinkle the quarters with salt and dry them in the oven at 145° F. for three to four hours until they are brittle. Store in any kind of airtight container and check periodically for mold. To rehydrate the dried sections when they are needed for cooking, pour about 1½ cups of boiling water over each cup of dried chips and let stand until all the water is absorbed, which takes about two hours. A third drying method is to use a small manufactured dehydrator such as Equi-Flow ($100).

To jar dried tomatoes in oil, let any small cherry-type tomatoes dry whole on the plants in the sun until they become leathery. Or dry them in the oven the same as above. Then insert a quarter clove of garlic, a basil leaf, and a pinch of ground pepper into each dried tomato. Pack the tomatoes together in a jar with olive oil, seal the jar, and store in a cool place. Check the jars occasionally to see if more oil has to be added. Use for appetizers after about four months' storage.

FREEZING
TOMATOES

Freezing is the simplest way to preserve tomatoes, but frozen tomatoes don't have the texture necessary for use in salads, being suitable for stews, soups, stewed tomatoes, and the like instead. Many experts advise that tomatoes can't be frozen—what they mean is that when thawed, the fruits yield more juice than pulp. Frozen tomatoes do have a fresh, uncooked flavor and will be good for up to a year.

Tomatoes are often frozen whole without any preparation at all. Perfectly ripe fruits are picked, wiped clean, and put in the freezer; when ready to be used, they're held under cold water and the skin peels right off. Others modify this method by scalding and removing the skin from the tomatoes first, and storing them in the freezer in heavy-duty freezer bags.

Another way to freeze tomatoes is to cut the fruits into slices or quarters, skin them, and pack them tightly in plastic containers with tight-fitting lids. Pack only what you'd use for one meal in each container, as thawed foods shouldn't be refrozen. The new square-shape freezer containers take up less room in the freezer.

Most tomato recipes can be frozen, too. Particularly good is frozen tomato juice. Just simmer a quart of trimmed and cored ripe fruits for ten minutes. Press these through a fine-mesh sieve and force the pulp through. Let the bowl cool in ice water and then pour the juice into plastic containers with tight-fitting lids, allowing plenty of space at the top. Do not add salt, pepper, or other spices until you thaw the frozen juice.

CANNING
TOMATOES

Although it is the best way to preserve tomatoes from taste and nutritional standpoints, canning is also the most difficult, expensive, risky, and time-consuming method. At the end of this chapter you'll find a list of tomato varieties especially good for canning, which can be added to from the variety lists in Appendixes I and II. Most of these types have thick flesh, fewer seeds, and are less juicy than other varieties. But be careful when using any canning variety. Episodes of botulism poisoning in the United States in 1975 exceeded the number of outbreaks in any year since 1935, and tomatoes, long considered a relatively safe food for canning, were involved in a great number of them. The problem seems mainly to be the fact that many of the new tomato varieties have a low-acid content, which supports the growth of botulism organisms more than high-acid fruits. *In fact, the USDA now recommends mixing citric acid with any tomatoes to be canned—one half teaspoon to the quart—to offset this prevalent low acidity.* It would be a good idea to do so, just as it would be to take a course in home canning. (Your state agricultural agent will advise where free courses are being held.) But at the very least follow all of these directions *carefully;* otherwise, the saints may be preserving *you:*

San Marzano, a large paste-type tomato, is often used for canning.

Choosing a Tomato-Water-Bath Canner. Organisms that cause food spoilage— molds, yeasts, and bacteria—are always present in the air, water, and soil. And enzymes that may cause undesirable changes in flavor, color, and texture are present in raw tomatoes and other vegetables. When you can tomatoes you must heat them hot enough and long enough to destroy spoilage organisms. This heating (or processing) also stops the action of enzymes. Processing of tomatoes is done in what is called a boiling-water-bath canner.

Water-bath canners are widely available on the market. You can also use any big metal container if it is deep enough for the water to be well over the tops of the tomato jars and to have enough space to boil freely. Allow two to four inches above the jar tops to insure brisk boiling. The canner must have a tight-fitting cover and a wire or wooden rack. If the rack has dividers, jars will not touch each other or fall against the sides of the canner during processing.

Choosing Glass Jars for Canning. Be sure all jars and closures are perfect. Discard any with cracks, chips, dents, or rust, for defects in jars prevent airtight seals.

Wash glass jars in hot, soapy water and rinse well. Wash and rinse all lids and bands. Metal lids with sealing compounds may need boiling or holding in boiling water for a few minutes. (Follow the manufacturer's directions.)

If you use rubber rings, use clean, *new* rings of the exact size for the jars. Don't test them by stretching. Wash rings in hot, soapy water. Rinse well.

Heinz 1706, another good tomato for paste.

Royal Chico, a more recent paste-tomato development.

Selecting Tomatoes for canning. About 2½–3½ pounds of tomatoes will make a quart of canned tomatoes. Choose fresh, firm fruits and can them before they lose their freshness. If you must store them, keep them in a cool, airy place.

For best quality in the canned product, use only perfect tomatoes. Sort them for size and ripeness; they cook more evenly that way. Do not use overripe tomatoes—they lose acidity as they mature.

Washing Tomatoes for Canning. Wash all fruits thoroughly—dirt contains one of the hardest-to-kill bacteria. Wash small lots at a time under running water or through several changes of water (lifting the tomatoes out of the water each time so that dirt that has been washed off won't return). Rinse the pan thoroughly between washings. Don't let the fruits soak (they may lose flavor and food value) and handle them gently to avoid bruising.

Processing and Packing Canning Tomatoes. Tomatoes may be packed raw into glass jars, or they can be preheated and packed hot in the jars—the so-called cold-pack and hot-pack methods. To loosen tomato skins for either method, dip the fruits into boiling water for about half a minute, then dip quickly into cold water. Cut out stem ends and peel the tomatoes.

1) For a raw or *cold pack* of processed tomatoes, pack the peeled, uncooked tomatoes to a half inch of the top of the glass jar, pressing gently to fill all spaces. Add no water. Add a half teaspoon of salt to pint jars; one teaspoon to quart jars. Adjust the closures carefully. Closures for glass jars are of two main types:

a) Metal Screw Band and Flat Metal Lid with Sealing Compound: To use this type, wipe the jar rim clean after the tomatoes are packed. Put the lid on, with the sealing compound next to the glass. Screw the metal band down tight by hand. When the band is tight, this lid has enough give to let air escape during processing. Do not tighten the screw band further *after* taking the jar from the canner when processing is finished. Screw bands that are in good condition may be reused another time, but the metal lids with sealing compound should be used only once.

b) Porcelain-lined Zinc Cap with Shoulder Rubber Ring: When using this type, first fit the wet rubber ring down on the jar shoulder, but don't stretch it unnecessarily. Fill the jar with tomatoes and wipe the rubber ring and jar rim clean. Then screw the cap down firmly and turn it back a quarter inch. As soon as you take the jar from the canner when processing is finished, screw the cap down tight, to complete the seal. Porcelain-lined zinc screw caps may be reused as long as they are in good condition. The rubber rings should not be reused.

When the lids on either type jar for canning tomatoes are adjusted properly, process the filled jars for the cold-pack method. Put them into a water-bath canner containing hot but not boiling water. Add boiling water if needed to bring the water an inch or two over the tops of the jars, but don't pour boiling water directly on the glass jars. Then put the cover on the canner. When water in the canner comes to a rolling boil, start to count processing time. Boil gently and steadily, processing pint jars of tomatoes thirty-five minutes and quart jars forty-five minutes. Add boiling water during processing, if needed, to keep the jars covered. Remove the jars from the canner immediately when processing time is up. Follow times carefully. If you live at an altitude of a thousand feet or more, you must add to the processing time as follows:

1,000 feet	2 minutes
2,000 "	4 "
3,000 "	6 "
4,000 "	8 "
5,000 "	10 "
6,000 "	12 "
7,000 "	14 "
8,000 "	16 "
9,000 "	18 "
10,000 "	20 "

2) For a *hot pack* of processed tomatoes, quarter peeled tomatoes as for the cold pack above *and cook them first, just bringing them to a boil and stirring to keep the tomatoes from sticking.* Pack and process these boiling hot tomatoes the same as for the cold pack, except that both pint and quart jars of the hot pack should be boiled for only ten minutes.

3) For processed *tomato juice,* wash and remove the stem ends from the tomatoes and cut them into pieces. Simmer until softened, stirring often. Then put them through a strainer. Add one teaspoon of salt to each quart of juice. Reheat at once just to boiling. Then use the same packing and processing method as for the hot pack above.

Cooling Canned Tomatoes. As you take jars from the canner, complete the seals if you have used the porcelain-lined zinc screw-cap type jar described above. (Do *not* tighten the screw band with metal-lid type any further.) If any liquid boiled out from the jars in processing, do not open the jars to add more.

Cool the jars top side up. Give each jar enough room to let air get at all sides. Never set a hot jar on a cold surface; instead set the jars on a rack or on a folded cloth. Keep hot jars away from drafts, but don't slow the cooling process by covering them.

Day-After Canning Jobs. Test the seal on glass jars with porcelain-lined screw caps by turning each jar partly over in your hands. To test a jar that has a screw top and flat metal lid, press the center of lid; if the lid is down and will not move, the jar is sealed. Or tap the center of the lid with a spoon. A clean, ringing sound means a good seal. Store jars without leaks and check for spoilage before use.

Should you find a leaky jar, use the unspoiled tomatoes right away. Or can them again, emptying the jar and packing and processing the tomatoes as if they were fresh.

When jars are thoroughly cool, take off the screw bands from both types carefully. If a band sticks, covering it for a moment with a hot damp cloth will help loosen it. Wash all screw bands and store them in a dry place for reuse. Before storing the canned tomatoes, wipe all jars clean. Label the jars to show contents and date.

One of the most nutritious tomatoes, Caro-Rich is excellent for canning because its high vitamin content makes up for vitamins lost in processing.

Campbell's 28, like most Campbell varieties, is a good canner with conventional tomato shape.

Storing Canned Tomatoes. Properly canned tomatoes stored in a cool, dry place will retain good eating quality for a year. Canned tomatoes stored in a warm place near hot pipes, a range, or a furnace, or indirect sunlight may lose some of their eating quality in a few weeks or months, depending on the temperature. Don't store jarred tomatoes in damp places, either—dampness may corrode metal lids and cause leakage so the tomatoes will spoil. Excessive cold presents problems, too. To protect against freezing in an unheated storage place, it is a good idea to protect canned tomatoes by wrapping the jars in paper or covering them with a blanket.

Signs of Spoilage in Canned Tomatoes. Don't use canned tomatoes that show any sign of spoilage. Look closely at each jar before opening it. Bulging jar lids, or bulging rings, or a leak—these may mean that the seal has broken and the food has spoiled. When you open a jar, look for other signs, too—spurting liquid, an off odor, or mold.

It's also possible for canned tomatoes to contain the poison causing botulism *without* showing signs of spoilage. Therefore, to avoid any risk of poisoning, it is essential that the canner be in perfect order and that every canning recommendation outlined here be followed exactly.

As an added precaution, bring home-canned tomatoes to a rolling boil before using; then cover and boil for at least ten minutes. If the food looks spoiled, foams, or has an off odor during heating, destroy it. Burn the spoiled tomatoes or dispose of them so that they won't be eaten by humans or animals.

THE TOP CANNING TOMATOES

Campbell's 17	Jubilee	Red Cherry
Campbell's 1327	Long Red	Red Pear
Caro-Rich*	Manitoba	Red Top
Cherry	Marglobe	Roma
Chico	Nova	Rutgers
Crimson Giant	Parker	San Marzano
Doublerich*	Pear	Spring Boy
ES24	Pink Gourmet	Tuckqueen
Hybrid Pickle	Pink Slipper	Veeroma
Hybrid Spring Giant	Ponderosa	Yellow Pear

* Their high vitamin C content makes up for some of the vitamin C lost during the canning process.

Chapter Eighteen

MAKING IT: A LOVE APPLE COOKBOOK

Someone once wrote of the tomato, "Doubtless God could have made a better berry, but doubtless God never did." People all over the world would applaud these sentiments. Tomatoes are, in fact, such an integral part of Continental cuisine that it's hard to imagine a French, Italian, or Spanish meal without them. Here in America, it's much the same; literally thousands of tomato-based recipes are enjoyed by Americans. A Philadelphia woman who died in 1913 actually went so far as to leave a delicious tomato recipe as her last will. Written on one page in a handwritten book of kitchen recipes, under the title "Chili Sauce Without Working," this tasty testament was accepted by the probate court:

"4 quarts of ripe tomatoes, 4 small onions, 4 green peppers, 2 teacups sugar, 2 quarts of cider vinegar, 2 ounces ground allspice, 2 ounces cloves, 2 ounces cinnamon, 12 teaspoons salt. Chop tomatoes, onions, and peppers fine, add the rest mixed together and bottle cold. Measure tomatoes when peeled. In case I die before my husband I leave everything to him."

Following are some of our favorite tomato dishes, though we still feel that a garden tomato tastes best eaten warm from the vine with no more seasoning than a little salt. In cooking these acid fruits try to use enameled, Teflon, or stainless steel cookware to avoid tasting unpleasant chemical reactions. Where a recipe calls for skinning a tomato, just hold it under hot water from a faucet a few minutes, then hold it under cold water and peel.

BEEFY TOMATOES WITH BASIL

Slice sun-ripe beefsteak tomatoes and alternate them in overlapping rows with sliced mozzarella cheese. Season with pepper, salt, oil, vinegar, and minced fresh basil.

Another All-America favorite—the tomato salad.

ORGANIC TOMATO-VEGETABLE COCKTAIL

2 cups tomato juice
½ teaspoon salt
2 tablespoons lemon juice
1 teaspoon Worcestershire sauce
3 sprigs parsley

1 small stalk celery, cut in 2-inch pieces
1 small carrot cut in 2-inch pieces
1 small onion
4 ice cubes

Put all ingredients except ice cubes in a blender container. Cover and blend until vegetables are liquefied. Add ice cubes, one at a time, while blender is running and allow them to become liquefied.

COUNTRY TOMATO SOUP

5 ripe tomatoes
½ peeled onion
⅛ cup shredded green pepper
⅛ cup chopped celery
1 slice peeled lemon

¼ cup minced parsley
1½ teaspoons salt
2 whole cloves
1 bay leaf
½ teaspoon peppercorns

Scald, skin, and quarter tomatoes and place in pot with 2 cups of water and other ingredients. Cook over medium heat, stirring often, until mixture comes to a rapid boil. Then lower heat and simmer for 20 minutes more. Strain out the cloves, peppercorns, and bay leaf and finally put the soup through the blender. Reheat before serving. Serves four amply.

GAZPACHO

5 ripe tomatoes
1 whole pimento, drained
½ onion
½ green pepper
½ cup tomato juice
3 cloves garlic
5 cups tomato juice
¼ cup olive oil

⅓ cup wine vinegar
½ teaspoon red-hot pepper sauce
2½ teaspoons salt
½ teaspoon pepper
5 chopped ripe tomatoes
1 chopped cucumber
½ cup sliced scallions
1 chopped green pepper

Purée the first six ingredients in a blender and put in a large bowl. Then add remaining ingredients, stir well, and chill. Serve cold as soup. Yields 2½ quarts.

CAPTAIN SAM'S CLAM CHOWDER

Whether clam chowder should be made with tomatoes or milk is a dispute that will never be settled. Actually, milk was used before tomatoes, but the very first clam chowder was made with neither. Breton sailors shipwrecked on the Maine coast invented the dish by adding crackers, salt pork, potatoes, and other ship's stores they could salvage, to clams found on the beach and cooking it all in an iron pot called a *chaudière*. Anyway, whether your favorite is a milk or tomato base, you'll call for more of this old recipe given to us by a retired sea captain:

2 dozen clams
½ pound salt pork, cubed
1 bunch celery, chopped
4 onions, chopped
1 leek
salt and pepper to taste

2 packs carrots, chopped
6 potatoes, largely cubed
4 tomatoes, chopped
1 can tomato purée
1 tablespoon thyme leaves
½ pint muscatel wine

Steam clams in one cup of salted water only until they open. Discard shells and put clams and strained broth to one side. Cube salt pork and sauté in a large saucepan. When brown, add 2 cups water, celery, onions, leek, and salt and pepper. Cover and bring to a boil. Simmer for 10 minutes. Add sliced carrots, simmer another 10 minutes. Add potatoes and 1 cup of water and cook until all is tender, about 10 minutes, depending on size of vegetables. In the meantime cut the clams into chunks. When the potatoes are done, add peeled, chopped fresh tomatoes, cover and cook 15 minutes. Then add tomato purée, thyme, clams, and clam broth. Cook 5 minutes and add the muscatel. Turn off and let sit for 1 hour before heating and serving. Serves twelve.

UNUSUAL TOMATO HEALTH SALADS

Try growing some rocket (also called ruca, roquette, rugola, and arugula) next year. Combined with tomatoes and cucumber as a salad green this mildly pungent herb makes for a salad that needs no further seasoning save a little oil and minced garlic.

Other unusual greens to substitute for lettuce in tomato salads include spinach, Good King Henry (often called "hardy spinach") the leaves of *Cichoria Catalogna* (asparagus chicory), and the young leaves of the French vegetable *mâche*.

ROASTED TOMATOES

2 cups cherry tomatoes
2 tablespoons olive oil
salt

pepper
¼ cup Parmesan cheese
2 tablespoons chopped parsley

Coat the tomatoes with the olive oil. Sprinkle with salt and pepper, and bake in a shallow dish 10–12 minutes at 400° F. Sprinkle with cheese and parsley, and serve. Serves four.

HERBED CHERRY TOMATOES

1 pint cherry tomatoes
2 tablespoons oil
1 clove garlic, crushed
2 tablespoons chopped parsley

1 tablespoon chopped basil
1 teaspoon salt
1 teaspoon chopped dill
freshly ground pepper

Plunge tomatoes in hot and then cold water to peel. Heat oil in pan, adding garlic and cooking for 1 minute. Then add peeled tomatoes and cook, stirring for 15 minutes. Sprinkle on herbs, cook 2 minutes more, and serve hot. Serves four.

SHORT-CUT STEWED TOMATOES

Stewed tomatoes can be made with fresh fruits, but this recipe is easier and tastes just as good. Just pour a large jar (or store-bought can) of whole tomatoes packed in tomato purée in a 3-quart saucepan. Break up the whole tomatoes with a fork. Add 2 tablespoons butter, 1 tablespoon sugar, ¼ teaspoon salt, and ⅛ teaspoon pepper. Stir and simmer on a low flame for about 10 minutes. Add 2 slices of fresh white bread broken into pieces and continue cooking until hot. Diced sautéed green pepper may also be added, according to taste. Serves six.

FRIED GREEN TOMATOES*

Nothing was wasted on U.S. farms of old—not even tomatoes that failed to ripen before the frost, as this recipe shows. Simply slice green or firm, almost ripe tomatoes into thick slices. Dip the slices in flour and fry each in a little fat or cooking oil until both sides are brown, pouring off excess tomato liquid while frying. The results are delicious.

* For anyone interested, the unusual, highly-specialized *Green Tomato Cookbook* ($3.45) is available from Writers, Inc., 1820 Terry Avenue, Seattle, Washington 98101.

EASY STUFFED TOMATOES

American cooks have stuffed tomatoes with almost every ingredient available and, as a seasoning, basil has traditionally been used here to bring out the best in any tomato dish. For excellent, easy stuffed tomatoes, scoop the insides from 4 ripe beefsteaks (or use the special hollow stuffing tomatoes mentioned in Chapter Three). Chop the pulp and drain the liquid. Combine this with ½ chopped onion, ½ cup bread crumbs, 1 tablespoon of chopped parsley, 3 tablespoons of grated Parmesan cheese, and 1 teaspoon of basil. Heap the mixture back into the shells and dot the tops with butter. Bake in a moderate oven 25 minutes until tops are brown. Serve warm or cool. Serves four.

STUFFED TOMATOES À LA SICILIENNE

Hollow out 6 tomatoes. Fill with a mixture of ⅛ teaspoon thyme, 1 tablespoon of fresh chopped basil, ¾ cup of ground boiled ham, 1 chopped onion, and 1 tablespoon of Madeira—binding this mixture first with mayonnaise thinned with milk. Sprinkle with bread crumbs dotted with butter and bake at 375° F. for about 25 minutes. Serves six.

STUFFED TOMATOES À LA PROVENÇALE

Follow the same method as for Stuffed Tomatoes à la Sicilienne, except use the following mixture for a stuffing: 1 clove of garlic crushed, ¾ cup of chopped mushrooms sautéed in butter, ½ cup large white bread crumbs fried in the mushroom butter, salt and pepper, and a dash of grated nutmeg.

STUFFED TIPPED TOMATOES

Remove skins from firm tomatoes by dipping in boiling water. Cut in halves, salt to taste, and arrange cut side up on a baking dish. Chop ham and green olives together and arrange in a mound on top of each tomato. Sprinkle a little olive oil and basil over the tomatoes and broil until hot and brown. Before serving add a pitted cherry or olive to the top of each hot tomato.

TOMATO SURPRISE!

Peel 2 large tomatoes and hollow them out very slightly, placing in a baking dish. Take stalks of firm canned asparagus and brush with a sauce made of butter and a little basil. Insert one asparagus stalk in each slightly hollowed tomato. Sprinkle with Parmesan cheese, pouring melted butter over this, and bake in a hot oven for about 20 minutes.

TOMATO OMELETTE

Cut 2 large ripe tomatoes into eighths. Melt butter in a large frying pan and stir in ½ cup cooked leftover shredded ham or chicken. Add tomato eighths to the meat for a moment, add more butter, and then pour 2 slightly beaten eggs over both. Sprinkle 1 tablespoon chopped parsley over the top, cover, and cook over a low flame to your taste. Serves four.

ITALIAN TOMATO SAUCE MAGGIORE

3 tablespoons olive oil	1 tablespoon salt
1 clove chopped garlic	½ teaspoon pepper
1 pound ground beef	6 Italian sweet sausages
1 large can Italian tomatoes	2 Italian hot sausages
¾ can water	4 chicken wings (optional)
1 large can Italian tomato paste	2 pork chops (optional)
1 teaspoon oregano	

Pour olive oil into an 8-quart saucepan. Add chopped garlic and lightly brown. Add ground beef and brown well. Pour in tomatoes and break up with a fork. Fill empty tomato can ¾ full with cold water, add to sauce. Bring to a slow boil and add the tomato paste, and seasonings. Stir well. Lower heat so that the sauce simmers gently. Add browned sausages (and chicken wings and pork chops if you like), stirring well. Continue simmering for 2 hours. Skim off any oil that rises to the top. Serves eight.

TWO WAYS TO MAKE CANDIED TOMATOES

1) Peel about 4 pounds of tomatoes for this old favorite. Mix 2 cups sugar and 4 cups water, adding a 2-inch stick of cinnamon to the mixture and bringing to a boil. Add the syrup to the tomatoes and cool slowly for 45 minutes or until the dish thickens. Remove cinnamon and serve hot. Serves six.

2) Another version is to combine 4 cups quartered peeled tomatoes, ¼ cup chopped onion, 6 tablespoons brown sugar, and fresh ground black pepper in a heavy skillet. Cook this on a low heat, uncovered, until the moisture is almost gone. Then cover with buttered bread crumbs and bake at 375° F. until brown. Serves six.

TOMATO CATSUP

Cook 8 quarts (¼ bushel) of ripe, washed, unpeeled tomatoes—this should leave a gallon of stock. Let stand for several days and then force through a wire sieve. Add ½ pint vinegar, 2 cups sugar, ⅓ cup salt, 3½ teaspoons cinnamon, ⅛ teaspoon cayenne pepper. Bottle and refrigerate for use. Makes about four quarts.

FIVE-MINUTE TOMATO FRENCH DRESSING

¼ cup lemon juice
1 cup salad oil
½ cup vinegar
1 cup tomato catsup
1 small piece onion
1 clove garlic
½ cup sugar
1 teaspoon salt
¾ teaspoon dry mustard
½ teaspoon paprika

Add all ingredients, in order listed, to a blender container. Blend on speed 9 until smooth. Makes approximately a quart.

CHILI SAUCE

Scald, skin, and chop 8 quarts ripe tomatoes. Add to these 1 cup chopped green peppers, 1 cup chopped onions, ½ cup salt, 5 cups sugar, 1½ quarts vinegar, 2 teaspoons cloves, 3 teaspoons cinnamon, 2 teaspoons ginger, and 2 nutmegs. Boil 3 hours. Makes about six quarts.

CHUTNEY

6 green tomatoes
4 onions
2 red peppers, seeded
1 cup raisins
1 quart vinegar

2 cups brown sugar
2 tablespoons dark mustard
2 tablespoons salt
2 tablespoons celery seed
12 sour apples

Cut vegetables in small pieces. Add raisins, vinegar, sugar, and spices and cook slowly for 1 hour. Core and quarter apples, add to mix, and cook slowly until sauce is soft and thick. Yields 4 pints.

PICCALILLI

2 quarts green tomatoes
5 pounds cabbage
6 sweet peppers
4 onions, medium-sized
1 cup salt

3 pints white vinegar
⅔ cup sugar
½ pound mustard seed
2 tablespoons celery seed
1 tablespoon horse-radish

Chop and mix tomatoes, cabbage, sweet peppers, and onion. Add salt and let stand 12 hours. Mix vinegar, sugar, mustard seed, celery seed, and horse-radish in a large pot. Bring to a boil. Add vegetables and heat until they are warm. Makes about 5 quarts.

GREEN TOMATO MINCEMEAT

Cook the following for 2¼ hours over a medium heat: 8 quarts cored green tomatoes, 3 pounds brown sugar, 2 pounds seeded raisins, 1 cup vinegar, 1 pint boiled cider, 1 tablespoon cloves, 1 tablespoon cinnamon, 1 tablespoon nutmeg, 1 tablespoon allspice, 1 teaspoon salt. Makes 5 quarts.

YELLOW TOMATO CONSERVE

Combine 2 quarts of small yellow tomatoes (unpeeled and unsliced) with 1 lemon cut in thin slices, 8 cups sugar, 1 cup water, and ½ pound chopped candied ginger (optional). Cook until thick. Makes 8 pints.

GREEN TOMATO MARMALADE

Wash 4 pounds of green tomatoes, trim, and cut into small pieces or slices. You will need 2 pounds sugar, ½ teaspoon salt, and 5 lemons. Peel lemons and cut into thin slices, and boil for 5 minutes in 1 cup of water. Slice the lemon pulp and remove the seeds. Combine the tomatoes, sugar, salt, sliced lemon, and drained peel. Heat slowly until sugar is dissolved. Boil 1 hour until the mixture is thick and the fruit clear.

BLOODY MARY

It's said that this drink was invented at the famous Harry's Bar in Paris in 1920 and named by an anonymous American expatriate who proposed that it be called the Bloody Mary because it reminded him of the Bucket of Blood Club in Chicago and he had a girl named Mary. It has since become famous as a hangover "remedy," the foremost example of the bit-of-the-hair-of-the-dog-that-bit-you school:

8 ounces tomato juice
3 ounces vodka
6 tablespoons lemon juice
1 egg white
½ teaspoon salt

dash of freshly ground pepper
2 fresh celery leaves
4 dashes of Worcestershire sauce
1 cup cracked ice

Put all ingredients in a blender and run until the ice is completely blended. Serve cold.

TOMATO WINE

This recipe is from a nineteenth-century cookbook and has not been tested. But it's the only tomato wine recipe we could find, so if you want something to carry you through winter until you can plant next year's tomato patch:

"Pick small, ripe tomatoes off the stems, put them in a clean bucket or tub, mash well, and strain through a linen rag (a bushel will make five gallons of juice). Add from two and a half to three pounds of brown sugar to each gallon. Put in a cask and let it ferment. If two gallons of water be added to a bushel of tomatoes, the wine will be as good. . . ."

Bon appétit!

Appendix I

OVER ONE HUNDRED
EARLY-RIPENING TOMATO VARIETIES
(bear fruit in forty-five to sixty-five days from transplanting outside)

No longer are early tomatoes all small and poor in quality. Though they aren't quite as tasty on the whole as main-crop types, many early varieties are delicious, medium-sized, and have other valuable qualities. There are far more *determinate* and *semideterminate* types that need no staking among the early varieties. Early varieties not only get garden-fresh tomatoes on the table sooner, but they are particularly valuable for sections of the country with short growing seasons. You'll find many types here suitable for your region and the diseases and/or special conditions prevalent there. These tomatoes are red unless otherwise noted and seed sources are given where the variety is hard to come by:

Atom—65 days—small to medium-sized fruits on a prolific plant that is good for growing indoors winters in a sunny window.

Bonny Best—65 days—medium-sized fruit on a vine adapted to short growing seasons in the North that has been a favorite since before World War One.

Burgess Early Salad—45 days—a small cherry tomato variety six to eight inches high with a two-foot spread; the earliest of all tomatoes, each plant yielding up to three hundred fruits.

Burpeeana Early Hybrid—58 days—a very heavy yielder whose fruits average 5 ounces; good for greenhouses.

Burpee's Big Early Hybrid—62 days—a prolific yielder of fruits that are said to average 7.6 ounces but can be much smaller.

Burpee's Globemaster Hybrid—65 days—tasty fruits averaging 7 ounces and very free of cracking.

Burpee's Gloriana—55 days—a standard plant yielding abundant fruits averaging about 5 ounces.

Burpee's Sunnybrook Earliana—58 days—another standard plant producing 5-ounce fruits.

Bush Beefsteak—61 days—produces medium to large meaty fruits on a bushy hybrid plant.

Calmart—60 days—medium fruits on a bushy plant resistant to fusarium and verticillium wilts and nematodes; adaptable to semiarid climates.

Coldset—65 days—medium-sized (4 ounces) tomatoes on a plant bred for the North that will set fruit well in cool soil; good for direct seeding outdoors.

Currant—63 days—small tomatoes shaped like currants on a bushy plant.

Doublerich—65 days—medium-sized fruits rich in vitamins on a plant with good disease resistance that bears until frost; has sixty units of vitamin C compared to twelve to twenty-five in other tomatoes.

Droplet—65 days—small, sweet fruits on a determinate vine that needs no staking.

Dwarf Champion—65 days—the best *pink* midget tomato, the fruits in clusters on a bushy plant; available from Stokes Seeds.

Earliana—see *Burpee's Sunnybrook Earliana*

Earliest of All—60 days—a small to medium-sized fruit that is actually fifteen days later than some varieties.

Early Twilley Big Hybrid—65 days—medium-sized fruits on a bushy hybrid plant resistant to fusarium and verticillium wilts.

Early Bird—57 days—a medium-sized fruit (averaging 6.4 ounces) that bears prolifically.

Early Delicious—55 days—fruits range from medium to large in size.

Early Girl—52 days—medium-sized tomatoes on a vine resistant to verticillium wilt.

Early Girl Hybrid—54 days—a heavy cropper yielding fruits in the 4–5 ounce range.

Early Red Cherry—56 days—yields small 1-ounce fruits good for eating out of hand or in salads.

Early Red Chief—65 days—a hybrid plant producing 1.2-ounce cherry tomatoes.

Early Stokesdale ✳4—64 days—small to medium-sized fruits on a plant developed for Canada and other northern areas.

Early Wonder—65 days—yields large fruits averaging 7.8 ounces.

Faribo Springtime—59 days—a prolific cold-resistant type whose small fruits are a bit soft inside.

Faribo Hybrid EE—55 days—a little earlier than the above and thus even better for cold areas.

Fireball—64 days—a determinate (non-staking) type with 4.6-ounce fruits on a plant adapted to the East and North and resistant to verticillium and fusarium wilts; good also for taking slips in August for winter indoor plants.

Firesteel—65 days—medium-sized, crack-resistant fruits on a plant resistant to fusarium wilt and blossom-drop in dry weather.

Fordhook Hybrid—60 days—a prolific yielder of uniformly shaped fruits averaging 6 ounces; especially good for sections with short summers.

Galaxy—65 days—a determinate (non-staking) type good for the North; is resistant to verticillium wilt.

Gardener—59 days—yields a medium-sized fruit resistant to cracking.

Gardener's Delight—65 days—a standard vine that produces fruits with a "delicious old-fashioned tangy flavor" that won a prize last year at Great Britain's Royal Horticultural Society trials; available from Thompson & Morgan.

Geneva—60 days—a fine cold-resistant type bred especially for the North.

German Dwarf Bush Imp—46 days—an improved form of the old German *Dwarf Bush;* small compact plant for pot or outdoor culture that resists 28° temperatures, yielding fruits two inches or larger.

Golden Delight—60 days—yields yellow-orange fruit, medium-sized, with excellent low-acid flavor.

Gloriana—see **Burpee's Gloriana**

Grape Tomato—56 days—really a cherry tomato, the plants bearing large clusters of little red tomatoes.

Hybrid Red ⚡22—58 days—medium- to large-sized fruits with good flavor.

Hybrid Pickle—60 days—fruit round, red, firm, the size of a crabapple—good for pickling.

Hybrid Pink ⚡1—63 days—the earliest hybrid pink tomato, 5–6-ounce fruits.

Hy-Top—64 days—an indeterminate plant producing tasty, medium-sized fruits.

Jetfire—60 days—a determinate (non-staking) type resistant to fusarium and verticillium wilts.

Johnny Jumpup—50 days—fruits the size of walnuts on a plant that is said to produce *all its crop in one hundred days from planting the seed.*

Mandarin Cross—65 days—the earliest F1 hybrid; orange-yellow, 9-ounce fruit on a spreading plant that must be staked; available from Gleckers Seedsmen.

Manitoba—60 days—a canning type good for the North.

Maritimer—59 days—medium-sized fruits that are light green when fully ripe; much used for canning.

Marmande—65 days—a typical Continental-type tomato available from Sutton's; irregular shape, very little seed.

Mixed Ornamental—65 days—a British tomato from Sutton's for outdoors and the greenhouse; bears a mixture of red and yellow fruits in currant, pear, and plum shapes on the same plant.

New Yorker—65 days—a determinate (non-staking) plant bearing 5.2-ounce fruits and resistant to verticillium and fusarium wilts; sets fruit in cool weather. Stokes' Special Transplanting New Yorker strain is particularly good.

Outdoor Girl—58 days—medium-sized fruit on a vine that grows about 4 feet high; available from Thompson & Morgan.

Park's Extra Early Hybrid—65 days—medium-sized fruit on a variety well-adapted to forcing in the greenhouse.

Patio—65 days—an extremely attractive hybrid, determinate (non-staking) variety that is, however, staked when grown in large pots on the patio; yields cherry tomatoes on a plant twenty-four to thirty inches high that is resistant to fusarium wilt.

Paul Bunyan—59 days—large 10–14-ounce fruits on a plant developed for cold northern sections that is resistant to sunscald.

Pixie Hybrid—52 days—cherry-type fruits averaging 1¾ inches on and eighteen-inch-high determinate (non-staking) plant with a twenty-four-inch spread; very attractive in six- to eight-inch pots or in baskets.

Porter—56 days—a determinate (non-staking) plant yielding cherry tomatoes that was bred for the Southwest and is drought resistant.

Presto—60 days—small cherry tomatoes on a thirty-inch hybrid plant that looks attractive and does well in six- to eight-inch pots.

Pretty Patio—50 days—similar to Patio, but earlier.

Primabel—65 days—determinate (non-staking) French type that produces smooth, round, medium-sized fruits on a bushy fifteen-inch plant.

Pusa Ruby—60 days—a new variety from India available from Gleckers Seedsmen; small red fruits on a disease-resistant vine.

Quebec ✳314—60 days—medium-sized fruits on a plant bred for the cold, cold North.

Queens Certified—60 days—medium to large fruits on a cold-resistant plant that is also resistant to fusarium wilt; one of the largest earlies.

Red Cloud—62 days—bred for Texas, this large-fruited plant has built-in drought resistance.

Red Knight—60 days—medium-sized, crack-resistant fruits on a plant resistant to many diseases.

Red Sugar—60 days—small (0.4 ounce) fruits on a plant that grows fairly tall.

Rideau—64 days—4–6-ounce fruits on a semideterminate plant that needs little staking.

Rocket—50 days—a determinate (non-staking) type bred for the North and bearing small cherry tomatoes.

Scotia—60 days—medium-sized fruits on a vine that does quite well in cool regions.

Selandia—60 days—medium-sized fruits on a vigorous plant; should be staked.

Sioux—63 days—an early cold-resistant type with medium-sized fruit.

Sleaford Abundance—52 days—an early ripener that yields large fruits on a plant that grows about eighteen inches high; available from Thompson & Morgan.

Small Fry—55 days—this All-America winner does well throughout the country, the hybrid determinate (non-staking) plant growing about three feet high and yielding up to 100 one-inch-diameter cherry tomatoes; a favorite for patio planters, Small Fry is resistant to fusarium and verticillium wilts and nematodes.

Spring Giant Hybrid—67 days—another All-America winner; yields fairly large (7–8-ounce) fruits on a determinate (non-staking) plant resistant to verticillium and fusarium wilts.

Springset—65 days—a rugged hybrid plant that is determinate (non-staking), Springset produces medium-sized fruit and is resistant to verticillium and fusarium wilts and nematodes.

Starfire—56 days—6-ounce fruits on a plant that performs well in light sandy soils.

Starshot—55 days—a new variety that averages twenty-five red but low-acid fruits weighing about 3 ounces each on a plant resistant to verticillium wilt.

Stokes Alaska—55 days—cherry tomatoes on a plant that does well in tubs and is good for northern sections.

Stokes Early Hybrid—56 days—small to medium-sized fruits on a vigorous plant that should be staked.

Subarctic—45 days—a very early, small-fruited plant that can be seeded directly into the garden; *Subarctic Medi* and *Subarctic Plenty* have larger fruits and bear slightly later.

Sugar Lump—65 days—sweet, cherry type tomatoes of about one inch in diameter on a vigorous plant that is cold-resistant.

Sugar Red—65 days—one-and a-half-inch fruits said to be sweeter than most cherry types.

Sugar Yellow—65 days—same as *Sugar Red* above except for color.

Summer Sunrise—60 days—6-ounce fruits on an indeterminate plant that should be staked.

Sunset—65 days—5–7-ounce tomatoes from a bushy vine that protects fruits from sunscald.

Surecrop—65 days—a determinate (non-staking) type plant resistant to late blight that yields medium-sized fruits.

Sutton's Alicante—62 days—an offering of Sutton's in England that is said to be especially disease resistant.

Swift—54 days—a bushy, determinate (non-staking) plant yielding medium-sized fruits.

Tanana—55 days—3–4-ounce tomatoes on a determinate plant bred for Alaska and parts north that is resistant to sunscald.

Tangella—65 days—from Sutton's in England; medium-sized fruits of intense tangerine color on a standard plant.

Tigerella—65 days—one of the best novelties; small delicious fruits colored red with golden stripes on a vigorous disease-resistant plant; available only from Sutton's.

Tiger Tom—65 days—also available only from Sutton's, these small red fruits have attractive stripes in red or orange-yellow.

Tiny Tim—55 days—probably the world's smallest tomato plant at twelve inches or so when fully grown; yields three-quarter-inch cherry-type fruits on a standard, determinate (non-staking) plant that is ideal for six to eight inch pots, window boxes, baskets, and for planting in borders.

Tom Tom—58 days—medium-sized fruits on a vigorous hybrid plant that is resistant to verticillium and fusarium wilts.

Valiant—65 days—medium-sized fruits on a vigorous, standard plant.

Veemore—60 days—a bushy plant that is fusarium and verticillium-wilt resistant and protects fruits against cracking.

Vigor Boy—63 days—6-ounce fruits on a plant resistant to verticillium and fusarium wilts and nematodes.

Vivid—60 days—an early pink type good for greenhouses; similar to *Earliest of All*.

Vogue—58 days—good-sized red fruits on a vigorous plant that needs staking.

Wayahead—60 days—a cold-resistant variety yielding medium-sized fruits.

Window Box—63 days—a cherry type suitable for large pots and window boxes.

Yellow Tiny Tim—55 days—much like *Tiny Tim* save for color.

Appendix II

MAIN-SEASON AND LATE TOMATOES
(bear in sixty-six to one hundred days)

Main-crop tomatoes are usually large, hybrid indeterminate (tall-growing) plants that must be staked and can be pruned without harming the plant or lessening yield. They bear the largest fruits by far and their flavor generally beats that of the earlies. The same applies to late tomatoes, varieties bearing in roughly eighty-one to a hundred days, a few of which should be planted to extend the tomato growing season into autumn. The really large main-crop and late tomatoes are difficult to grow in sections of the country with short growing seasons, but are a must everywhere else. All of these tomatoes are red unless otherwise noted, and both hybrid plants and determinate (non-staking types) are noted. Late crop varieties are asterisked:

A&C Pole Boy—77 days—extra-large fruits on a vigorous, tall-growing plant.

Abraham Lincoln—71 days—fruits large, often over a pound; a very decorative plant with attractive bronze foliage; available from Shumway.

Ace—70 days—medium-sized fruits on a plant adapted to the West and resistant to verticillium and fusarium wilts. Ace Improved, 55, Cal-Ace, Cal-J, and Ace Improved Royal are also widely available.

African Beefsteak—see *Pink Ponderosa.*

Anahee—72 days—a Hawaiian-bred plant resistant to the spotted-wilt virus common there and elsewhere.

Apollo—74 days—an indeterminate, hybrid plant that yields large fruits and is resistant to fusarium and verticillium wilts and nematodes.

Atkinson—80 days—a southern-bred determinate plant resistant to fusarium and verticillium wilt that yields fairly large tomatoes and does well in humid climates.

Avalanche—68 days—medium to large fruits on a hybrid plant resistant to fusarium wilt that sets fruit in spite of drought and heat.

Basket Pak—76 days—bite-sized, 1½-inch cherry tomatoes on a prolific vine.

Bay State—70 days—a small-fruited plant resistant to the leaf-mold fungus common in greenhouses and found in some sections outdoors.

Beefeater—see *Beefmaster Hybrid* below.

*Beefmaster Hybrid—75–82 days—very large (1½–2 pounds) deep pink fruits on a plant resistant to verticillium and fusarium wilts and nematodes. Formerly called *Beefeater*.

*Beefsteak—82 days—an old favorite; very large, ribbed fruits low in acidity on a standard plant that is also called *Crimson Cushion*.

Better Boy Hybrid—72 days—large (1½ pounds) fruits on prolific plants resistant to verticillium and fusarium wilts and nematodes.

*Big Crop Climbing—90 days—one of the latest tomatoes; giant fruits on vines that grow fifteen feet and more.

Big Johnny—80 days—a hybrid plant yielding large red fruits.

Big Red—80 days—an F1 hybrid plant that yields large crimson-red fruits.

Bigset—76 days—large fruits on a high-yielding hybrid plant that is bushy in habit and resistant to fusarium and verticillium wilts and nematodes.

Big Seven—77 days—big globular fruits produced in abundance by a hybrid vine resistant to verticillium and fusarium wilts and nematodes.

*Boatman Miracle Climbing Tomato—90 days—a grower reported 221 fruits on *one* of these plants, which climbs up to sixteen feet high.

Bonanza—75 days—a bushy determinate (non-staking type) developed in South Dakota that yields large-sized fruits.

Bonus—75 days—fairly large fruits on a bushy plant that affords protection against sunscald.

Boone—75 days—a variety that has proved resistant to fusarium wilt over the years and produces medium-sized fruits.

Bradley—72 days—pink, medium-sized fruit on a plant resistant to fusarium wilt that does well in humid climates.

Break O'Day—70 days—medium to large fruits on a cold-resistant plant that is also resistant to verticillium and fusarium wilts.

*Brimmer—90 days—giant meaty tomatoes up to 3 pounds and low in acid; particularly good in the South.

*Burgess Jumbo Hybrid—81 days—mild fruits up to 2 pounds on a medium-tall vine with good foliage cover.

Burgess Stuffing Tomato—78 days—almost hollow, seedless fruits that are ready to stuff when you pick them; the tomatoes grow 3¼ inches across and 2¾ inches deep.

Burpee Big Boy Giant Hybrid—78 days—large 1- to 2-pound tomatoes on a popular plant that does well in all but very humid areas and has dense foliage that protects fruits from sunscald.

Burpee's Delicious—77 days—a standard variety producing tasty fruits up to 2 pounds that are resistant to cracking.

Burpee's Early Girl VF—78 days—a new introduction; *Burpee Big Boy's* kid "sister," which yields fruits up to a pound on a vigorous hybrid plant that is resistant to fusarium and verticillium wilts.

Burpee's Globe—80 days—medium-sized pink tomatoes in clusters of six to ten on a standard variety that is an All-America winner.

Burpee's Jubilee—72 days—another All-America winner—with large, orange fruits rich in vitamin C.

Burpee's Matchless—78 days—a heavy cropper, this standard variety yields medium to large, smooth, long-keeping fruits.

Burpee's Table Talk—75 days—solid, meaty 7–8½-ounce fruits on a standard plant.

Burpee's VF Hybrid—72 days—medium-to-large fruits on a heavily foliaged plant resistant to verticillium and fusarium wilts and cracking.

California—73 days—good-sized fruit on a disease-resistant vine appropriate for the Southwest.

Campbell's 17—75 days—a semideterminate (non-staking) paste tomato good for canning; resistant to growth cracks and gray-leaf spot.

Campbell's 1327—75 days—a semideterminate (non-staking) canning type developed for the North that yields 8.6-ounce fruits and is resistant to verticillium and fusarium wilts.

Cardinal Hybrid—74 days—medium-sized, crack-resistant fruits on a vigorous indeterminate plant.

Caro-Red—75 days—a nutritious orange-red variety developed in the Midwest that has a normal vitamin C content but ten times the provitamin A of standard red varieties; available from Stokes Seeds.

Cherry—72 days—small, scarlet cherry tomatoes good for salads, preserves and eating out of hand.

Chico—75 days—a determinate (non-staking) variety resistant to fusarium wilt that yields small tomatoes good for tomato paste; *Chico Grande* has slightly larger fruits.

***Climbing Trip-L-Crop**—90 days—large fruits 2 pounds and more on a vine that climbs up to twenty feet high.

Colossal—85 days—giant fruits up to 2½ pounds on a large, indeterminate hybrid plant.

Crackproof—78 days—medium-sized, scarlet fruits resistant to fruit cracking.

Crackproof Pink—78 days—a pink tomato resistant to fruit cracking.

Creole—72 days—a new high-yielding southern variety resistant to fusarium wilt and blossom-end drop.

Crimson Cushion—see *Beefsteak*.

*Crimson Giant—90 days—very large fruits of up to 1½ pounds on a vigorous vine.

Crimson Sprinter—70 days—medium-sized, highly crimson tomatoes, low in acid, on a determinate plant that doesn't need to be staked.

Cura—70 days—a hybrid, disease-resistant plant producing large crops of medium-sized fruit; available from Thompson & Morgan.

Dutchman—74 days—large purple-pink fruits that are low in acid; an old-fashioned variety, seeds available only from Gleckers.

Earlibell—70 days—small to medium-sized fruit of good quality.

Early Detroit—78 days—a medium-sized pink tomato on a reliable plant.

Early Giant—70 days—bigger fruits than on most plants bearing this early.

*Early Pack No. 7—81 days—medium-sized fruits on a disease-resistant vine developed for the Southwest.

Early Pink—74 days—a miniature much used for patio gardens and window boxes.

Egg Tomato—70 days—smooth-skinned, long-lasting, egg-shaped red fruits on a productive plant; fruits low in acid.

Epoch Dwarf Bush—70 days—good for window sills; fruits are larger than the earlier *Tiny Tim*.

ES24—75 days—a paste tomato with resistance to the gray-leaf spot common in the Southeast.

Essar—72 days—average-sized fruits on a plant resistant to fusarium and verticillium wilts.

Evergreen—70 days—light green when fully ripe (dark green when ripening), this tomato is low in acid and good for eating out of hand; available from Gleckers Seedsmen.

Fantastic—70 days—a hybrid with crack-resistant, medium-sized fruits on a vigorous plant that does well in the Midwest and East.

Firesteel—70 days—medium-sized fruits earlier than most main croppers on a prolific bushy vine with lots of foliage.

Florabel—75 days—a southern variety resistant to early blight, fusarium wilt, gray-leaf spot, and growth cracks.

Floralou—75 days—another southern-bred variety with dense foliage; resistant to growth cracks and gray-leaf spot.

Floramerica—75 days—a new All-America selection with fruits up to 1 pound on a plant resistant to fifteen tomato diseases, including verticillium and fusarium wilts.

Foremost E-21—70 days—fairly large tomatoes on a prolific plant good to seed directly into the garden.

German Sugar—72 days—somewhat resistant to the spotted-wilt virus.

*Giant Belgium—85 days—huge pink fruits up to 2 pounds often used in making tomato dessert wine because of its sweetness and low-acid content.

*Giant Oxheart—87 days—very large, 2-pound, non-acid fruits on a plant bearing late in the season.

*Giant Tree—90 days—reselected by an amateur gardener in the Chicago area, this standard Italian potato-leaved variety bears low-acid fruits up to 1½ pounds on a ten to twelve foot vine.

Glamour 77—a vigorous plant yielding large crack-resistant fruits; developed for the Midwest and Northwest.

Globelle—75 days—large, pink fruits on a plant resistant to leaf mold.

Garden State—70 days—a very productive vigorous plant bearing medium to large fruits.

Golden Boy—75 days—a sport of *Burpee Big Boy,* this hybrid plant is resistant to fusarium wilt and bears large yellow-orange fruits; its vitamin A content is lower by a third than standard reds, while its vitamin C content is about the same.

*Golden Oxheart—87 days—a golden-orange tomato; see *Giant Oxheart,* the plants similar except for fruit color.

Golden Ponderosa—75 days—a large-subacid, yellow-orange fruit; resembles *Ponderosa,* save for color.

Golden Queen—75 days—similar to *Golden Ponderosa* above.

Golden Sphere—75 days—another yellow-orange variety that is resistant to fusarium wilt.

Golden Sunrise—78 days—tangy golden-yellow fruits on a plant that is also good for the greenhouse; available from Thompson & Morgan.

*Greater Baltimore—82 days—a large red tomato much used for canning.

Gulf States Market—80 days—although it is usually grown commercially, this variety, bearing large purplish-pink tomatoes, is also recommended for the southern home garden.

Hawaii—73 days—one of a number of varieties developed in Hawaii to combat gray-leaf spot.

Heinz 1350—75 days—a standard semideterminate (non-staking) type adapted to the East and Midwest and good for canning; resistant to verticillium and fusarium wilts and cracking. *Heinz 1370, Heinz 1409,* and *Heinz 1419* are similar, varying mostly in shape.

*Henry Field Tomato—100 days—a new hybrid yielding large clusters of ½–¾-pound fruits on a strong-growing, disease-resistant vine.

Highlander—68 days—medium fruits on a small, determinate (non-staking) plant resistant to fusarium and verticillium wilts.

Homestead 24—76 days—a southern standard variety resistant to fusarium wilt and cracking.

Hotset—75 days—resistant to blossom drop caused by hot, dry weather in the Southwest and elsewhere.

Hybrid Spring Giant—72 days—a vigorous plant yielding a large tomato good for canning.

Hybrid Surprise—68 days—large fruit on a productive plant highly resistant to wilt diseases.

Hy-X-67—a short, bushy determinate (non-staking) vine that is resistant to many diseases and yields abundant crops; available from Henry Field.

Indian River—75 days—a southern variety resistant to growth cracks, gray-leaf spot, and graywall.

Italian Plum—75 days—small, plum-shaped fruit mostly used for tomato paste.

Jefferson—72 days—medium-sized fruit on a plant resistant to fusarium wilt.

Jetstar—72 days—a northern hybrid that yields fruits averaging 10.4 ounces on a vigorous plant resistant to verticillium and fusarium wilt.

Jubilee—see *Burpee's Jubilee.*

June Pink—69 days—a purplish-pink variety yielding medium-sized fruits low in acid.

*Jung's Giant Climber—87 days—large fruits on a vine that climbs over fifteen feet high.

Kaloki—75 days—another Hawaiian type resistant to spotted wilt.

Kokomo—75 days—a Hawaiian variety resistant to fusarium wilt.

*Kurihara 90—a large-fruited, pink Japanese variety adapted to American growing conditions that is becoming very popular.

*Lakeland Climbing Tomato—87 days—this vine grows small for a climber (six feet) but its fruits are said to be up to 3 pounds each.

Laketa—75 days—purple-pink, pear-shaped fruits with solid, sweet, non-acid flesh.

Liberty Bell—78 days—a bicentennial year introduction offered by Gleckers: "a bullnose 4-lobe stuffing tomato identical to the common bell pepper; thick shell walls and a small central seed core that is easily trimmed out; low in acid."

Livingston Globe—75 days—a vigorous plant yielding medium-sized pink fruits.

Long Red—76 days—good for canning, this tomato takes its name from its habit of bearing a long time.

Louisiana Wilt—75 days—resistant to fusarium wilt as its name implies.

Manahill 80—a variety developed in Florida for resistance to early blight, fusarium wilt, and gray-leaf spot.

Manalu—80 days—a southern type resistant to gray-leaf spot.

*****Manalucie**—87 days—bred in Florida, where diseases are active all year long, this widely grown old favorite is relatively free of growth cracks and resistant to verticillium and fusarium wilts, early blight, gray-leaf spot, graywall, and leaf mold.

*****Manapal**—85 days—a large-fruited southern but widely adapted variety relatively free of growth cracks and resistant to fusarium wilt, gray-leaf spot, leaf mold, amd blossom-end rot.

Manastota—76 days—a variety with built-in resistance to fusarium wilt.

Marbon—68 days—medium fruits on plant adapted to a wide range of growing conditions.

Marglobe—79 days—large tomatoes on a prolific plant resistant to fusarium wilt and nailhead spot; slips easily taken in August to root for indoor winter tomatoes.

Marion—76 days—medium to large crack-resistant fruits on a standard plant developed for the Southwest and resistant to gray-leaf spot.

Marvana—72 days—medium-sized fruits on a plant resistant to verticillium wilt.

Mars—70 days—a processing type bred for resistance to gray-leaf spot.

Maui—75 days—a variety developed for Hawaii and resistant to gray-leaf spot.

McGee—73 days—large fruits on a plant bred especially for the Southwest.

McMullen—74 days—a prolific plant that yields medium-sized pink tomatoes.

Merit—68 days—cherry-type fruits on a plant resistant to verticillium and fusarium wilts and nematodes.

Michigan-Ohio—75 days—hybrid plant developed for greenhouses that yields large fruits.

Mission Dyke—70 days—this disease and drought-resistant vine, which yields medium-sized fruits, was tested in Puerto Rico and "withstood all the tropics could give it"; available from Gleckers.

Monte Carlo—75 days—large, green-shouldered fruits on a hybrid plant resistant to verticillium and fusarium wilts and nematodes.

Moon Glow—72 days—medium-sized, orange fruit low in acid that keeps a long time.

Moira—70 days—5–6-ounce, highly crimson fruits with very few seeds.

Morden—70 days—a plant bred in Canada that yields the earliest of yellow-orange tomatoes.

Moreton Hybrid—70 days—medium-sized to large fruits on a popular plant most suitable for Canada, the Northeast and the East.

Morning Star—72 days—a new hybrid plant resistant to verticillium and fusarium wilts that yields fruits averaging 8 ounces.

Moscow VR—75 days—a verticillium-resistant vine that yields large fruits.

Motored—75 days—a variety bred for resistance to tobacco-mosaic virus.

Nemared—75 days—a southern plant bearing medium-sized fruits and resistant to nematodes and fusarium wilt.

Nepal—80 days—large tomatoes on a prolific, disease-resistant plant that hails from the Himalaya Mountains; available from Farmer Seeds.

New Hampshire—70 days—another variety developed for late-blight resistance.

No. 670—75 days—a wilt-resistant plant that tolerates both extreme wet and extreme dry conditions, its scarlet red fruits resistant to cracks and sunburn; available from Gleckers.

Nova—65 days—a paste tomato on a plant resistant to late blight.

*Oh! Boy—82 days—fruits on this new hybrid introduction are large and crack resistant; the plant is indeterminate and resistant to fusarium and verticillium wilts; available from Henry Field.

Ohio—75 days—a hybrid variety highly resistant to leaf mold.

Ohio-Indiana—75 days—a large pink type on a hybrid plant suitable for greenhouses.

Ohio Pink M-R 13—75 days—a greenhouse tomato resistant to tobacco-mosaic virus.

Ohio W. R. Pink Globe—78 days—another pink forcing variety good for greenhouse growing.

Ohio W. R. Seven—75 days—a variety with built-in resistance to graywall.

Orange Queen—75 days—a beefsteak tomato, bright orange in color.

Owyhu—70 days—a variety resistant to the curly top or western yellow blight common in the West and Midwest.

*Oxheart—86 days—very large, pink non-acid fruits up to 2 pounds on a tall standard plant that is an old favorite.

Pan American—70 days—this offspring of a cross with the Peruvian variety *Red Currant* is resistant to fusarium wilt.

Parker—75 days—a paste-type tomato with great resistance to growth cracks.

Park's Whopper—73 days—large fruits up to four inches in diameter on a vigorous plant.

Payette—70 days—a variety resistant to curly top or western yellow blight of the West and Midwest.

Pear—70 days—small red fruits excellent for canning and preserves.

Pearl Harbor—72 days—a medium-sized tomato on a plant resistant to spotted wilt and bred for low, moist areas.

Pearson ✗9—72 days—fruits averaging 9.4 ounces on a hybrid, determinate (non-staking) plant that is resistant to verticillium and fusarium wilts.

Pearson Improved—72 days—similar to *Pearson ✗9* above but with fruit relatively free of growth cracks.

Pelican—70 days—large fruits on a vine developed at Louisiana State University that is resistant to nematodes and fusarium wilt.

Perfecta—75 days—big yields of large tomatoes on a plant resistant to many tomato diseases.

Peron—70 days—a plant so disease resistant that its breeders (Gleckers Seedsmen) call it the "Sprayless Tomato"; fruits are medium-sized and high in nutritive value.

Petoearly—70 days—medium-sized fruits on a determinate plant especially good for semiarid climates and resistant to verticillium and fusarium wilts.

Petogro—72 days—similar to the above but more resistant to wilt disease.

Petomech II—72 days—a "square" type tomato that can stay in the field a long time without rotting, the plant resistant to verticillium and fusarium wilts.

Pink Deal—77 days—medium-sized pink tomatoes on a plant that sets fruit well in hot, dry weather.

Pink Delight—77 days—crack-resistant 7-ounce fruits on a new indeterminate, hybrid plant that yields prolifically.

Pink Gourmet—73 days—a hybrid pink that's a good canner; also called Pink Grapefruit.

Pink Lady Hybrid—75 days—mild, pink fruit of medium-size on a plant bred for the Northeast.

*Pink Ponderosa—86 days—large, low-acid, purplish-pink fruits 24 ounces and more on a tall standard plant that is also called *African Beefsteak;* fruits usually of grotesque shapes, not smooth and uniform.

Pink Shipper—75—a firm pink that is often used for canning.

*Pink-skinned Jumbo—86 days—very large pink fruits frequently weighing 2 pounds and more.

Plainsman—70 days—5-ounce fruits on a small determinate (non-staking) plant with thick foliage that protects against sunburn; good in dry areas.

*Pondeheart—85 days—a new Japanese cross of *Oxheart* and *Ponderosa* that yields large, pink, non-acid fruits; seed available from Gleckers.

Potomac—68 days—a variety resistant to verticillium and fusarium wilts and nematodes.

Pritchard—70 days—medium-sized fruits on a standard All-American winner resistant to fusarium wilt and very prolific.

*Ramapo—85 days—very large fruit on a hybrid plant resistant to verticillium and fusarium wilts.

Red Champion—70 days—a very heavy bearer of medium-sized tomatoes; disease-resistant vine.

Red Cherry—72 days—a large plant bearing cherry-type fruits good for eating whole or preserving.

Red Chief VFN Hybrid—80 days—large fruits on a vigorous plant especially developed for the South that has excellent disease resistance.

Red Currant—72 days—a cherry type with built-in resistance to leaf mold.

Red Ensign—70 days—heavy crops of medium-sized fruit on a hybrid, disease-resistant plant; available from Thompson & Morgan.

Red Glow—72 days—a hybrid plant that is resistant to fusarium and verticillium wilts and nematodes and yields average-sized, ideally shaped fruits.

Red Peach—72 days—fruits shaped like little vases that grow in clusters on a large plant.

Red Pear—72 days—same as *Red Peach* above.

Red Plum—72 days—same as *Red Peach* except for shape.

Red Rock—68 days—a prolific, tough-skinned variety developed for eastern commercial growers.

Red Top—75 days—large plum fruits suitable for tomato paste.

Red Whoppa—70 days—tremendous crops averaging 15 pounds on a standard plant with good disease-resistance; available from Thompson & Morgan.

Riverside—75 days—a variety resistant to fusarium and verticillium wilts.

*Rockingham—85 days—6–8-ounce, deep-red fruits on a vigorous plant that is resistant to late blight.

Roma—76 days—medium-sized paste-type tomatoes on a determinate (non-staking) plant resistant to growth cracks and verticillium and fusarium wilts; also called *Catsup* and *Roma Red*.

Ruffled Tomato—75 days—an unusual yellow fruit shaped like an accordion; excellent for scooping out interior to make dessert containers when cut in half; available only from Gleckers.

Rushmore—70 days—fruits 6 ounces and more on a hybrid plant popular in the Midwest.

Rutgers—75 days—perhaps the most popular variety in America over the years; flavorful 7-ounce fruits on a semideterminate, standard plant that is resistant to fusarium and verticillium wilts. *Rutgers* may or may not be staked and is good for taking slips to root for indoor tomato plants.

Rutgers Hybrid—80 days—a more disease-resistant plant than *Rutgers,* with larger tomatoes; *Rutgers California Supreme* is also available for the West.

Russian Red—74 days—medium-sized fruit on a determinate plant that is more tolerant to low temperatures than most tomatoes.

Saladette—70 days—round 2-ounce fruits on small plants that are resistant to fusarium and verticillium wilts, nematodes, and blossom-end rot.

San Marzano—80 days—a rectangular, non-cracking, paste-type tomato; this high-yielding vigorous plant is easy to take slips from to root for winter tomatoes (though San Marzano's plants must be hand-pollinated indoors).

San Pablo—80 days—a paste-type tomato very similar to San Marzano.

Sausage—80 days—a novelty, "funtastic" variety that is shaped something like a sausage.

Snowball—78 days—a mild, pure-white tomato that contains less acid than any other variety; available from Henry Field Seeds.

Southland—70 days—a variety bred to resist early blight and fusarium wilt.

Spring Boy—72 days—medium-sized to large fruits very suitable for canning.

Square Tomato—80 days—Gleckers Seedsmen offers this novelty, which is square, though with rounded shoulders; jointless stems, no stems ever on picked fruit.

Stakeless—80 days—8-ounce fruits with comparatively few seeds on a determinate (non-staking), strong-stemmed plant resistant to fusarium wilt and early blight; this good pot-patio type with big potatolike leaves also resists sunscald.

*Stone—81 days—6½- to 7½-ounce fruits on a standard plant long a late-crop favorite.

Strain-A Globe—73 days—bred for resistance to graywall disease common in Florida and elsewhere.

Summerset—78 days—resistant to blossom drop common in hot, dry climates like the Southwest.

Sunray—75 days—large golden-orange fruits on a prolific standard plant that is resistant to fusarium wilt; but while vitamin C is the same as in standard reds, *Sunray* has one third less provitamin A than most reds do.

Supercross—72 days—heavy crops on a hybrid plant resistant to tobacco mosaic and other diseases; available from Thompson & Morgan.

Superman—78 days—very large tomatoes on a prolific plant that is becoming a great favorite.

Supermarket—80 days—large, delicious fruits that don't at all taste as if they come from the supermarket; resistant to fusarium wilt and gray-leaf spot.

Super Master—80 days—a heavy yielder of large-sized tomatoes.

Super Sioux—70 days—medium-sized fruits on a hybrid plant bred for the Midwest and North.

Supersonic—79 days—a tall hybrid plant yielding large fruit that does particularly well in the East and Midwest and is resistant to fusarium and verticillium wilts and nematodes.

Tamiami, PS—75 days—a new introduction especially good for southern areas; large fruits on a determinate plant highly resistant to verticillium and fusarium wilts.

Tecumseh—75 days—a canning type resistant to gray-leaf spot.

Tennessee Red—74 days—bred especially for Tennessee and resistant to fusarium wilt.

Terrific—73 days—10-ounce fruits on a tall plant resistant to fusarium and verticillium wilts and nematodes.

Texto—75 days—large tomatoes on a plant developed for Texas.

Thessaloniki—68 days—baseball-sized fruits on a disease-resistant plant developed in Greece; Gleckers claims this is "a rare tomato that refuses to rot when ripe—laid on a shelf it slowly dehydrates itself without bursting skin or letting the juice run out."

Tipton—74 days—a new variety resistant to fusarium wilt.

Tom Boy—80 days—a vigorous, indeterminate vine yielding large tomatoes.

Toy Boy—72 days—a new introduction featuring abundant Ping-Pong ball sized fruits on an attractive plant resistant to wilt diseases; good for indoor and outdoor planting, especially for hanging baskets—three Toy Boy plants can be grown in a single ten-inch basket.

Traveller—78 days—crack-resistant 6–7-ounce fruits on an indeterminate vine somewhat resistant to fusarium wilt that is good for humid climates.

Trellis 22—70 days—a tall-growing climbing variety yielding medium-sized fruits.

Trimson—72 days—this plant has the most intense red color of all tomatoes, two to three times that of common red tomatoes; the fruits are very low in acid, 5–6 ounces, and borne on a determinate vine that needs no staking; available from Johnny's Selected Seeds.

Tropic—80 days—medium-sized fruits on a southern variety resistant to verticillium and fusarium wilts, gray-leaf spot, and mosaic virus.

Tropic-Gro—82 days—a southern variety resistant to graywall, gray-leaf spot, and fusarium and verticillium wilts.

Tropic-Red—80 days—another southern type resistant to gray-leaf spot and graywall.

Tuckcross 520—70 days—good for forcing in greenhouses; resistant to leaf mold.

Tuckcross M—70 days—another greenhouse type resistant to leaf mold.

Tuckqueen—70 days—a good greenhouse variety that sets fruit at comparatively low temperatures and little light.

Tucker—73 days—a new variety resistant to fusarium wilt.

Ultra Boy—72 days—large fruits on a plant resistant to verticillium and fusarium wilts and nematodes.

Ultra Girl—70 days—medium-sized fruits on a vigorous plant resistant to verticillium and fusarium wilts and nematodes.

Valiant—66 days—medium-sized fruits with a mild pleasant flavor on a plant that does well in humid climates.

Veebrite—69 days—a northern-bred variety resistant to verticillium and fusarium wilts and catface malformation.

Veeroma—72 days—a disease-resistant paste-type tomato.

Vendor—68 days—resistant to tobacco mosaic virus and leaf mold, this variety is excellent for greenhouses.

Venus—75 days—medium-sized fruits on a plant resistant to fusarium and southern bacterial wilts and cracking.

Vetomold—72 days—a new type for the greenhouse, resistant to leaf mold.

VF tomatoes—all of the many varieties with VF as their prefix are extremely resistant to wilts; most, however, are processing tomatoes of one kind or another.

VF13L-34—75 days—medium-sized fruits on a disease-resistant vine bred for the Southwest.

VFN-8—72 days—a western variety with medium-sized fruits on a determinate plant resistant to nematodes and fusarium and verticillium wilts.

Viceroy—74 days—large tomatoes on a plant that sets fruit well in cool weather and is a favorite of growers for its easy picking.

Vineripe—70 days—medium-sized to large fruits on a tall, vigorous hybrid plant.

Walter—79 days—meaty, medium-sized fruits on a determinate (non-staking) plant that is resistant to fusarium wilt, gray-leaf spot, and fruit cracking.

Watermelon Beefsteak—75 days—large, pink fruits somewhat oblate in form (like a watermelon) on a standard plant that dates back at least a century; available only from Gleckers.

West Virginia—79 days—a variety bred in West Virginia that is resistant to late blight.

*White Beauty—84 days—one of the best novelties: an ivory-white tomato, very sweet and low in acid; fruits larger than *Snowball;* available from Burgess Seeds.

White Queen—similar to *White Beauty* above.

White Wonder—similar to *White Beauty* above.

Whopper—73 days—see *Park's Whopper.*

*Winsall—84 days—very large pink fruits on a tall-growing plant.

Wisconsin Chief—80 days—a semideterminate vine that doesn't need much staking and yields large fruits up to about a pound.

Wonder Boy—82 days—very large, green-shouldered fruits on a prolific hybrid plant much grown in the East, Midwest, and South.

Yellow Jumbo—80 days—large tomatoes on a standard plant that has been grown for over 150 years; available from Twilley Seeds.

Yellow Pear—70 days—small-fruited, pear-shaped fruits good for salads, preserves, and pickling.

Yellow Plum—70 days—same as *Yellow Pear* above, except for plum-shaped fruit.

Appendix III

DEVELOPING A TOMATO VARIETY JUST FOR YOU

by William J. Sandok, New York State Cooperative Agent

Although hundreds of tomato varieties are listed in this book, you may be interested in developing an entirely new variety with a flavor, texture, size, or acidity perfectly suited to your taste. Or perhaps, factors like cold resistance, disease resistance, color, and yield are of concern to you. Or you may simply want to be the only gardener in town with your own "personalized" tomatoes. Whatever the reason, creating a new tomato variety is a relatively simple matter.

To create a new tomato, you should be sure to start with non-hybrid parents, *not* hybrids, as hybrid plants are much more difficult to cross. (Many non-hybrid or "standard" varieties are listed in Appendixes I and II.)

By using non-hybrid parents even the inexperienced gardener can develop a new tomato without much trouble at all. The first rule is to be observant. Take careful notes of non-hybrid plants in the garden with the characteristics you desire. For example, if you are interested in developing a processing tomato, one that can be used for canning and making preserves, you will most likely want one with lots of pulp and very little juice. If, on the other hand, you are looking for a tomato that has extremely large size but sets during cold weather, you can easily select fruit from early-setting and -ripening plants during the summer. The observant gardener can take advantage of many such variations.

Once you have selected the plant and/or the fruit on that plant with your desirable characteristics, attach a label to the flower cluster or fruit cluster with a string. Observe these through the season and if they continue to have the characteristics you desire, pick the fruit when it is fully ripe. The fruit should be a deep red color and soft to the touch. When the fruit is picked, allow it to continue ripening until it becomes very soft. Then it should be transferred to a container such as a glass jar where the seeds and pulp can be mixed together and mashed in the jar and allowed to ferment. This fermentation is advantageous in seed development because it produces an acid that kills surface bacteria and some viruses. Many plant breeders use this method (or one of those mentioned in Chapter Sixteen) simply to be sure that they will not transfer diseases when the seed is planted the following season.

Once the pulp and seed have fermented for several days, the seeds can be easily separated and thoroughly washed. After the acidic pulp is removed from the seed, the seed should be immediately dried and stored in a dried condition until ready to use. To be sure the seed is viable before planting, a small amount can easily be germinated in a test. A simple way to do this is to put a moist towel on a dish and put several seeds on top of the towel, covering them with a piece of plastic to keep them from drying out. After a period of several days at room temperature, germination should be evident. The next growing season, selectively choose those plants or tomatoes showing the desired characteristics and repeat the process again. After several summers of careful selection, you should have plants producing tomatoes suited to your needs.

There is also another interesting method that the home gardener can use to develop a new tomato variety. That is by crossing two non-hybrid varieties. Before trying this somewhat more complex method, you should know that each tomato flower is "complete," i.e., has both male and female parts. Each flower has five or more male stamens (they produce the pollen) which form a yellow cone around the pistil (the female organ). When the anthers (the top part of the stamens) produce pollen, it is transferred to the pistil through wind action. Insects such as bees and flies are not needed. During the warm, dry part of the day, pollen can easily be seen by shaking a ready flower.

To cross non-hybrid varieties, just remove the stamens from one flower on a tomato plant. Remove the entire cone or capsule. In this way there will be no chance of not removing the anthers on the top parts of the stamens, which contain the pollen sacs. Only the long female pistil should remain on the flower.

Next, take pollen from *another* variety nearby by means of a small artist's paintbrush. Transfer the pollen to the top of the female pistil (the stigma) on the first flower. That's all you have to do. To assure that extraneous pollen doesn't reach your selected flower, remove any nearby blossoms. Also be sure to mark your treated blossom with a label so it can be easily identified as the season goes by. By using two different types of non-hybrid parents a great deal of variation can result in the progeny or offspring. From this variation, additional varieties can be developed quite easily, using the method described previously for selection and seed extraction. You'll soon have tomato varieties you can name after yourself!

Appendix IV

SIXTY TOMATO-SEED SOURCES BY REGION

North—New England
Asgrow Seed Co., P. O. Box 725, Orange, Connecticut 06477
Brecks of Boston, Breck Building, Boston, Massachusetts 02210
Comstock, Ferre & Co., 263 Main St., Wethersfield, Connecticut 06109
Charles Hart Seed Co., Main & Hart Streets, Wethersfield, Connecticut 06109
Johnny's Selected Seeds, North Dixmont, Maine 04932

East
Abbott & Cobb, 4744–46 Frankford Avenue, Philadelphia, Pennsylvania 19124
F. W. Bolgiano & Son Inc., 411 New York Avenue, Washington, D.C. 20002
Burnett Brothers, Inc., 92 Chambers Street, New York, New York 10007
W. Atlee Burpee Co., Philadelphia, Pennsylvania 19132
Carroll Gardens, Box 310, Westminster, Maryland 21157
The City Gardens, 437 Third Avenue, New York, New York 10016
Fredonia Seed Co., 183 East Main Street, Fredonia, New York 14063
Joseph Harris Co., Moreton Farm, 3670 Buffalo Road, Rochester, New York 14624
Herbst Brothers, Inc., 1000 North Main Street, Brewster, New York 10501
Hygrade Seed Co., 30 Water Street, Fredonia, New York 14063
J. H. Kiltgord Co., 7347 East Main Street, Lima, New York 14485
D. Landreth Seed Co., 2700 Wilmarco Avenue, Baltimore, Maryland 21223 (America's oldest seed company, 1784—Washington and Jefferson were customers.)
Natural Development Co., Bainbridge, Pennsylvania 17502 (untreated "organic" seed)
Page Seed Co., Greene, New York 13778

Seedway (formerly Robson Quality Seeds), Hall, New York 14463

Stokes Seeds, Box 548, Main Post Office, Buffalo, New York 14240

Thompson & Morgan, Ltd., P. O. Box 24, 401 Kennedy Boulevard, Somerdale, New Jersey 08083

Otis S. Twilley Seed Co., P. O. Box 1817, Salisbury, Md. 21801

South

Brawley Seed Co., P. O. Box 594, Mooresville, North Carolina 28115

George W. Park Seed Co., Greenwood, South Carolina 29646

H. G. Hastings Co., P. O. Box 4088, Atlanta, Georgia 30302

Hilltop Farm & Garden Center, Box 866, Cleveland, Texas 77327

Kilgore Seed Co., P. O. Box PP, Plant City, Florida 33566

Reuter Seed Co., Inc., 320 North Carrollton Avenue, New Orleans, Louisiana 70119

Wetsel Seed Co., Harrison, Virginia 22801

Midwest

W. Atlee Burpee Co., Clinton, Iowa 52732

Burgess Seed & Plant Co., Galesburg, Michigan 49053

Farmer Seed & Nursery Co., Fairbault, Minnesota 55021

Henry Field Seed & Nursery Co., Shenandoah, Iowa 51601

Germania Seed Co., 5952 North Milwaukee Avenue, Chicago, Illinois 60646

Gleckers Seedsmen, Metamora, Ohio 43540

Gurney Seed & Nursery Co., Yankton, South Dakota 57078

R. L. Holmes Seed Co., 2125 46th Street N.W., Canton, Ohio 44709

J. W. Jung Seed Co., Randolph, Wisconsin 53956

Letherman's Inc., 501 McKinley Avenue, N.W., Canton, Ohio 44702

Earl May Seed & Nursery Co., Shenandoah, Iowa 51601

Midwest Seed Growers, 505 Walnut Street, Kansas City, Missouri 64106

L. L. Olds Seed Co., P. O. Box 1069, Madison, Wisconsin 53701

Harvey E. Saier, Diamondale, Michigan 98201

R. H. Shumway Seedsman, 628 Cedar Street, Rockford, Illinois 61101

Spring Hill Nurseries, Tipp City, Ohio 45371

Sunnybrook Farms Nursery, 9448 Mayfield Road, Chesterland, Ohio 44026

Vaughn Seed Store, 5300 Katrine Street, Downers Grove, Illinois 60515

West

D. V. Burell Seed Growers, Inc., Box 150, Rocky Ford, Colorado 81067

W. Atlee Burpee Co., Riverside, California 92502

A. L. Castle, Inc., 190 Mast Street, Morgan Hill, California 95037

Desert Seed Co., P. O. Box 181, Centro, California 92243

Gill Brothers Seed Co., P. O. Box 16128, Midway Station, Portland, Oregon 97216

Keystone Seed Co., Box 1438, Hollister, California 95023

Nichols Garden Nursery, 1190 North Pacific Highway, Albany, Oregon 97321

Rogers Brothers Seed Co., P. O. Box 2188, Idaho Falls, Idaho 83401

Taylor's Garden, 2649 Stingle Avenue, Rosemead, California 91770

Canada

McFayden Seeds, P. O. Box 1600, Brandon, Manitoba, Canada

Overseas

Sutton & Sons Seed Growers, Ltd., Reading, England

Thompson & Morgan, Ltd., Ipswich, England

The photographs in this book are used through the courtesy of
the following companies:

George J. Ball, Inc., pages: 49, 50 bottom, 89, 135, 151 bottom left and top right.
Burgess Seed & Plant Co., pages: 10, 20, 24, 155.
W. Atlee Burpee Company, pages: 18, 30, 40, 87 left, 90, 136 bottom, 161 bottom.
Chapin Watermatics, page 109.
Farmer Seeds & Nursery Company, pages: 87 right, 91.
Ferry Morse and Company, pages: 15, 56, 57, 78, 79, 94, 136 top right, 180, 187.
John & Elaine Gittens Photography, page 8.
Joseph Harris Company, pages: 65 top, 85, 136 top left, 151 top left.
Hendrickson, Brian, page 82.
Herts & Brothers Seedsmen, page 27.
Lakeland Nurseries, page 23.
Lord & Burnham, page 170 top.
Northrup, King & Company, page 61.
Petoseed Company, Inc., pages: x, 17 right, 22, 28, 29, 64, 83 bottom, 159, 172, 181.
Status Seeds, pages: 14, 63, 65 bottom, 163 top, 170 bottom.
Suncrop, Inc., page 50 top.
Sutton & Sons Seed Growers, Ltd., pages: 21, 96, 109 right, 163 bottom, 167 top.
Tatuland Nurseries, pages: 17 left, 26, 156.
United States Department of Agriculture, pages: xvi, 11, 53, 54, 60, 69, 73, 83 top, 95,
112, 113, 129, 132, 133, 137, 138, 141, 142, 144, 146, 147, 148, 157, 160, 161
top, 168 top and bottom.

INDEX

Acidity
  soil, 34–35, 76
  tomato, 19, 180
  water, 108
African violet soil, 167
Agricultural Research Station (Texas), 63
Agrotronics Manufacturing Company, 109
Alkalinity
  soil, 34–35, 76
  water, 108
All-America tomato varieties, 29, 64
Allerton, Frank, *Ring Culture* by, 93
Aluminum foil mulch, 112, 114
American Nurserymen's Association, 29
Amiben herbicide, 111
Ammonium nitrate, 100
Anahee tomato variety, 147
Animal damage prevention, 124, 127, 128, 130, 131
Anthracnose, 97, 134, 136
Aphids, 125–26
Artificial light, 58, 167–69
Aztecs, 2

Bacillus thuringiensis, 123, 127, 133
Bacterial spot, 137
Bamboo and Rattan Works, 85
Bare-root plants, 41
Basic H insecticide, 124, 126
Beard, James, xi
Beefmaster tomato variety, 14, 135
Beefsteak tomato variety, 14, 20
Big Red starter kit, 50
Biotrol XK insecticide, 123
Birds, 121, 126
Bishop, Michael, "Rogue Tomato" by, 7
Black plastic mulch, 63, 83, 112, 113, 116
Blake College, 119
Blight control, 134–49
Blister beetles, 126
Bloody Mary recipe, 195
Blossom drop, 138
Blossom-end rot, 76, 138
Blossom-set spray, 70
Bordeau spray, 144
Border-plant tomato varieties, 26

Boron, 103, 104
Botulism, 180
BR8 peat blocks, 51
Brooklyn Botanic Garden, 172
Bug Bait, 123
Burgess Seed & Plant Company, 10, 12, 25, 27, 40
Burke, James A., xi
Burpee, W. Atlee, Company, 12, 59
Burpee's Automator variety, 80
Burpee's Big Boy variety, 12, 21, 72
Burpee's Globe variety, 19, 29
Burpee's Jubilee variety, 19, 29
Burpee's VF Hybrid, 136

Cal-ace variety, 28
Calcium, 103, 104
Calmart variety, 162
Campbell's Tomato Soup, 6
Campbell varieties, 28, 139, 184
Candied tomato recipes, 193
Cannibal's tomato, 10–11, 13
Cantharides, 126
Cape gooseberry, 25
Captain Sam's clam chowder, 188–89
Carbonated water, 107
Carême, Antonin, 3–4
Caro-Red variety, 18
Caro-Rich variety, 184
Carver, George Washington, 9
Catfacing, 139
Catsup (ketchup), 3, 4, 7, 193
Chapin Watermatics, 108, 109
Cherry tomatoes, 42, 190
Chicken Marengo, 3
Chili sauce recipes, 186, 193
Chinese tomato ring, 90, 154
Chlordane, 127
Chutney recipe, 194
Climbing varieties, 17, 87, 151
Cold frame, 66–68, 165–66
Coldset variety, 41, 42, 65
Colorado Agricultural Experiment Station, 115
Commentaries on the Six Books of Dioscorides (Mattioli), 3
Companion planting, 97–98, 123

Compost, 76, 102–3, 158
Container planting, 151, 154–64, 168
Controlling Tomato Diseases, 149
Copper, 103, 105
Cornell University, 71
Corset growing, 90
Cortez, Hernando, 2
Country tomato soup, 188
Cracking, 139
Cucumber mosaic, 139
Cultivation, 110
Cuttings, 66, 118
Cutworms, 80, 117, 123, 126–27
Cyphomadra betacea, 23

Damping off, 44, 140
DCPA herbicide, 111
DDT pesticide, 123
Defoliation, 97
Destructive and Useful Insects—Their Habits and Control (Metcalf and Flint), 121
Dipel insecticide, 123
Direct seeding, 41–42
Disease control, 120–21, 134–49
Disinfectants, soil, 45
Dixie Hybrid, 163
Doom insecticide, 124, 128
Dorrance, John, 6
Doublerich variety, 18, 185
Drainage, 37–38
Du Pont watering kit, 109
Duraset spray, 70

Early blight, 140–41
Early Girl variety, 89
Echinocystis lobate, 130
Eggplant-tomato cross, 26
Egg tomatoes, 22
Elizabeth I, 3
Entocons insecticide, 124
Equi-Flow dehydrator, 179
Eugénie, Empress, 3, 4
Euphorbia lathyris, 128
Evergreen variety, 20

Fantastic variety, 28

supersize, 13–15, 16
tangerine, 19
tree, 22–24, 154
wet area, 28
white, 19, 20
yellow, 18, 19
Tomato wine, 27, 196
Toy Boy variety, 159
Trap plants, 123, 128
Tree tomatoes, 22–24, 154
Trimson variety, 19
Trip-L-Crop variety, 87
Tropic variety, 143
Tumbling Tom variety, 151, 168
2, 4-D weed killer, 111, 139, 148, 149
2, 4-5-T weed killer, 149

Ultra Boy variety, 14
Ultra Girl VFN variety, 63
Under Glass magazine, 172
United States Department of Agriculture,
    xi, 11, 28, 32, 68, 107, 108, 111, 117,
    129, 131, 158, 167, 180
United States Supreme Court, 6

University of Illinois, 172
University of Toronto, 19
USDA Soilless Fertilized Mix, 167

Van Meeter, James, 5
Vermiculite, horticultural, 45–46
Verticillium wilt, 148, 149
VF 10 variety, 83
Virus diseases, 139, 140, 146, 147
Vitamin Institute, 105
Vitamins, 7, 18, 25, 105, 184, 185

Walnut wilt, 149
Warhol, Andy, 6
Washington, George, 4
Watering, 57, 59, 71, 73, 80, 106–10, 157,
    162, 167, 171
Watersaver Systems, 108
Waynesboro Nurseries, 89
Weed control, 110–11
Weed killers, 111, 139, 148, 149
When to plant, 52, 72, 74
Where to plant, 74–75, 150, 153, 154
    in containers, 151, 154–64, 168

in greenhouse, 169–72
indoors, artificial light, 167–69
indoors, sunny window, 27, 163, 166–67
in rented gardens, 152–53
to save space, 150, 153–54
White Beauty variety, 20
Wickson, Edward J., 13
Willow stakes, 88–89
Window boxes, 151, 158
Windsor, Conn., garden space, 152
Wire-cage technique, 90–92, 154
Wolf peach, 1, 4, 5
World Art and Gift, 92
Writers, Inc., 190

Xtomatl, 1

Yellow tomato conserve, 195

Zilke Brothers, 89
Zinc, 103, 104–5
Zineb dust, 134, 144